Journal of Applied Logics - IfCoLog Journal of Logics and their Applications

Volume 11, Number 3

June 2024

Disclaimer

Statements of fact and opinion in the articles in Journal of Applied Logics - IfCoLog Journal of Logics and their Applications (JALs-FLAP) are those of the respective authors and contributors and not of the JALs-FLAP. Neither College Publications nor the JALs-FLAP make any representation, express or implied, in respect of the accuracy of the material in this journal and cannot accept any legal responsibility or liability for any errors or omissions that may be made. The reader should make his/her own evaluation as to the appropriateness or otherwise of any experimental technique described.

© Individual authors and College Publications 2024
All rights reserved.

ISBN 978-1-84890-457-6
ISSN (E) 2631-9829
ISSN (P) 2631-9810

College Publications
Scientific Director: Dov Gabbay
Managing Director: Jane Spurr

http://www.collegepublications.co.uk

All rights reserved. No part of this publication may be used for commercial purposes or transmitted in modified form by any means, electronic, mechanical, photocopying, recording or otherwise without prior permission, in writing, from the publisher.

EDITORIAL BOARD

Editors-in-Chief
Dov M. Gabbay and Jörg Siekmann

Marcello D'Agostino	Michael Gabbay	David Pym
Natasha Alechina	Murdoch Gabbay	Ruy de Queiroz
Sandra Alves	Thomas F. Gordon	Ram Ramanujam
Jan Broersen	Wesley H. Holliday	Chrtian Retoré
Martin Caminada	Sara Kalvala	Ulrike Sattler
Balder ten Cate	Shalom Lappin	Jörg Siekmann
Agata Ciabattoni	Beishui Liao	Marija Slavkovik
Robin Cooper	David Makinson	Jane Spurr
Luis Farinas del Cerro	Réka Markovich	Kaile Su
Esther David	George Metcalfe	Leon van der Torre
Didier Dubois	Claudia Nalon	Yde Venema
PM Dung	Valeria de Paiva	Rineke Verbrugge
David Fernandez Duque	Jeff Paris	Jun Tao Wang
Jan van Eijck	David Pearce	Heinrich Wansing
Marcelo Falappa	Pavlos Peppas	Jef Wijsen
Amy Felty	Brigitte Pientka	John Woods
Eduaro Fermé	Elaine Pimentel	Michael Wooldridge
Melvin Fitting	Henri Prade	Anna Zamansky

Scope and Submissions

This journal considers submission in all areas of pure and applied logic, including:

- pure logical systems
- proof theory
- constructive logic
- categorical logic
- modal and temporal logic
- model theory
- recursion theory
- type theory
- nominal theory
- nonclassical logics
- nonmonotonic logic
- numerical and uncertainty reasoning
- logic and AI
- foundations of logic programming
- belief change/revision
- systems of knowledge and belief
- logics and semantics of programming
- specification and verification
- agent theory
- databases
- dynamic logic
- quantum logic
- algebraic logic
- logic and cognition
- probabilistic logic
- logic and networks
- neuro-logical systems
- complexity
- argumentation theory
- logic and computation
- logic and language
- logic engineering
- knowledge-based systems
- automated reasoning
- knowledge representation
- logic in hardware and VLSI
- natural language
- concurrent computation
- planning

This journal will also consider papers on the application of logic in other subject areas: philosophy, cognitive science, physics etc. provided they have some formal content.

Submissions should be sent to Jane Spurr (jane@janespurr.net) as a pdf file, preferably compiled in LaTeX using the IFCoLog class file.

CONTENTS

ARTICLES

Frontiers of Logic and Computation in China 255
Juntao Wang, Yanhong She, Pengfei He and Jiang Yang

Algebraic Study of Substructural Fuzzy Epistemic Logics 259
Yongwei Yang and Yijun Li

Algebras of Similarity Monadic Fuzzy Predicate Logic 279
Xuesong Fu, Xiaoyan Liu and Zhiqin Zhao

Monadic Operators on Bounded L-algebras 301
Lingling Mao, Xiaolong Xin and Xiaoguang Li

Ideals on Pseudo Equality Algebras 327
Zhaoping Lu and Xiaolong Xin

Some Types of Weak Hyper Filters in Hyper BE-algebras 355
Xiaoyun Cheng, Xiaoguang Li and Xiaoli Gao

Special Issue on Frontiers of Logic and Computation in China

JUNTAO WANG
School of Science, Xi'an Shiyou University, China
`wjt@xsyu.edu.cn`

YANHONG SHE
School of Science, Xi'an Shiyou University, China
`yanhongshe@xsyu.edu.cn`

PENGFEI HE
School of Mathematics and Statistics, Shaanxi Normal University, China
`pfhe@snnu.edu.cn`

JIANG YANG
School of Mathematics, Northwest University, China
`yangjiangdy@126.com`

Nowadays, logic has covered more and more aspects of natural and social science, from mathematics, physics and computer science to philosophy, cognitive science and linguistics, and has found application in virtually all aspects of information technology, from software engineering and hardware to programming and artificial intelligence. Indeed, logic, artificial intelligence, cognitive science and theoretical computing are influencing each other to the extent that a new interdisciplinary area of logic and computation is emerging.

In recent years, Chinese scholars have made a great effort to promote logical research and have achieved great success in the aspects of logic and computation, mainly include classical and non-classical logic, algebraic logic, modal and temporal logic, probabilistic logic, aggregation function and fuzzy implication, knowledge-based systems and knowledge representation, automated reasoning and so on.

This special issue aims to provide an opportunity for Chinese researchers to worldwide share their novel ideals, original research achievements, and practical experiences in a broad range of logic and computation. Topics include, but are certainly not limited to:

—Non-classical logic and Non-monotonic logic
—Algebraic logic
—Temporal logic and Dynamic logic
—Probabilistic logic and Fuzzy logic
—Aggregation function and Fuzzy implication
—Logic programming and Logic-based approaches
—Approximation reasoning and Automated reasoning
—Soft Computing and Granular Computing
—Knowledge-based systems and Knowledge representation.

This special issues features five contributions from areas described above.

The first contribution "Algebraic study of substructural fuzzy epistemic logics" by Yongwei Yang and Yijun Li, they generalize the notion of monadic residuated lattices to that of pseudo monadic residuated lattices, which serve as algebraic models of modal logic **KD45(FL$_e$w** and discuss the relationship between pseudo monadic residuated lattices and other pseudo monadic algebraic structures, showing that it is a natural generalization of pseudo monadic BL-algebras, Bi-modal Gödel algebras and pseudo monadic algebras. They also provide a comprehensive characterization of pseudo monadic residuated lattices by considering them as pairs of residuated lattices (L, B), where B represents a special case of a relatively complete subalgebra of L known as c-relatively complete.

The second contribution "Algebras of similarity monadic fuzzy predicate logic" by Xuesong Fu, Xiaoyan Liu and Zhiqin Zhao, they introduce the notion of similarity monadic MTL-algebras and give some representation of this algebras based on filters. They also construct the logic of the variety of similarity monadic MTL-algebras and prove the (chain) completeness of this logic.

The third contribution "Monadic operators on bounded L-algebras" by Lingling Mao, Xiaolong Xin and Xiaoguang Li, they introduce the notion monadic bounded L-algebras as L-algebras equipped with two monadic operators $\forall$ and $\exists$. They also discuss the relation between monadic bounded L-algebras and monadic quantum B-algebras and any other monadic algebras. These results are important to the further algebraic study of related logic systems with monadic operators.

The forth contribution "Ideals on pseudo equality algebras" by Zhaoping Lu and Xiaolong Xin, they introduce kinds of ideals on pseudo equality algebras, which are possible algebraic semantics of higher fuzzy logic, and provide some characterizations of them.

The fifth contribution "Some types of weak hyper filters in hyper BE-algebras", they introduce the notion of weak hyper filters in hyper BE-algebra and study including positive implicative weak hyper filters, implicative weak hyper filters and obstinate weak hyper filters. The authors also discuss the relations between (positive) implicative weak hyper filters and weak hyper filters (obstinate weak hyper filters and maximal weak hyper filters, positive implicative hyper filters) respectively. They also give some equivalent characterizations of these weak hyper filters under some certain conditions.

The editors are grateful to all the authors, and equally to the reviewers, for their contribution. Special thanks go to Dov M. Gabbay and Jane Spurr for giving some excellent suggestions for improving this special issue.

Algebraic Study of Substructural Fuzzy Epistemic Logics

Yongwei Yang
School of Mathematics and Statistics, Anyang Normal University, China
yangyw@aynu.edu.cn

Yijun Li*
School of Financial Mathematics and Statistics, Guangdong University of Finance, China
liyijun2009@126.com

Abstract

This paper generalizes the notion of monadic residuated lattices to that of pseudo monadic residuated lattices. As monadic residuated lattices serve as algebraic models of modal logic $\mathbf{S5(FL_{ew})}$, we propose pseudo monadic residuated lattices as algebraic models of modal system $\mathbf{KD45(FL_{ew})}$. The main contributions of this paper are as follows: 1) we discuss the relationship between pseudo monadic residuated lattices and other pseudo monadic algebraic structures, showing that it is a natural generalization of pseudo monadic BL-algebras, Bi-modal Gödel algebras and pseudo monadic algebras; 2) We provide a comprehensive characterization of pseudo monadic residuated lattices by considering them as pairs of residuated lattices (L, B), where B represents a special case of a relatively complete subalgebra of L known as c-relatively complete. Furthermore, we establish a necessary and sufficient condition for a subalgebra to be c-relatively complete.

keyword: residuated lattice; epistemic logics, pseudo monadic residuated lattices, relatively complete

---

The authors would like to express their grateful to anonymous reviewers for their comments and effort. The works described in this paper are partially supported by Henan Provincial Soft Science Research Projects for Development of Science and Technology (No. 232400410084, 222400410052), Guangdong Province Philosophy and Social Sciences Planning 2023 Youth Project (No. GD23YGL17) and Characteristic Innovation Project of Guangdong Province Department of Education (No. 2022WTSCX078).

*Corresponding author.

1 Introduction

Non-classical logic is more suitable for handling uncertain and fuzzy information compared to classical logic. In the past several decade years, numerous fuzzy logical algebras have been proposed as the semantic units for non-classical logic systems. For example, MV-algebras were introduced in [1] by Chang as algebraic models of the infinitely-valued logic of Łukasiewicz, while BL-algebras were introduced in [2] by Hájek as algebraic semantics of basic fuzzy logic, a general framework in which tautologies of continuous t-norm and their residua can be captured [3]. Inspired by Hájek's work, Esteva and Godo proposed a new formal deductive system monoidal t-norm based logic in [4], and intended to cope with left-continuous t-norms and their residua [5]. However, all the above mentioned algebras are the particular cases of residuated lattices, which were introduced by Dilworth in [6] and stemmed from attempts to generalize properties of the lattice of ideals of a ring, so residuated lattices are very basic and important algebraic structures. The study of filter theory is crucial in investigating the subdirect representation theorem of fuzzy logical algebras and establishing the completeness of their corresponding logical systems. From a logical perspective, various filters naturally interpret as sets of provable formulas. Recent studies on residuated lattices have explored different types of filters [7].

Epistemic logics have been proposed to provide explicit insights into knowledge and belief [8]. However, human practical reasoning demands more than what traditional classical epistemic logic can offer. Classically, the truth of a statement q with respect to a state of knowledge K is determined if every model of K is also model of q. But nothing can be said about its truth value if only the most possible models of K are also models of q. The situation becomes even more complex when we need to acknowledge that the statement q can have an intermediate truth value different from true. The typical semantics for fuzzy epistemic logic is Kripke-style semantics. To address this, Hájek proposed a fuzzy possibilistic semantics for a system of epistemic logic. Unfortunately, finding an axiomatization of the underlying possibilistic logic of **BL** based on this semantics is not straightforward because both K and C axioms are not valid (i.e., distributivity of necessity over $\to$ and $*$, respectively). To overcome this problem, Busaniche et al. approached it in a novel way by proposing a possible algebraic semantics, which is obtained by extending BL-algebras (the algebraic models of basic logic) with two operators that model necessity and possibility.

So far, the only many-valued extensions of minimal logics axiomatized in the literature are the ones corresponding either to a finite Heyting algebra ([9, 10]), or to the standard (infinite) Gödel algebra [11] or to a finite residuated algebra [12] (in particular finite Łukasiewicz linearly ordered algebras). In the present paper, we aim to provide an algebraic study of a generalization of fuzzy epistemic logic, namely substructural fuzzy epistemic logic.

The rest of the paper is structured as follows: in order to make the paper as self-contained as possible, in Section 2, we provide a recapitulation of the fundamental concepts related to substructural fuzzy logics and their algebraic semantics, which will be used in this paper. In Section 3, we introduce the notion of substructural fuzzy epistemic logic $\mathbf{KD45}(\mathbf{FL_{ew}})$ and its algebraic semantics pseudo monadic residuated lattices. In Section 4, we demonstrate that pseudo monadic residuated lattices generalize three well studied classes of algebras, including pseudo moandic BL-algebras, Bi-modal Gödel algebras and pseudo monadic algebras. Section 5 presents some construction methods of pseudo monadic residuated lattices. In Section 6, we conclude the paper with final considerations, discuss future work, and highlight potential further applications.

2 Preliminaries

In this section, we will provide a summary of some results regarding the substructural logic $\mathbf{FL_{ew}}$, which refers to the full Lambek calculus with exchange and weakening, as well as its algebraic semantics, namely, residuated lattices. These concepts will be utilized in the context of this paper.

Definition 2.1. [2] $\mathbf{FL_{ew}}$ consists of the following axioms and rules:
(1) $(\alpha \Rightarrow \beta) \Rightarrow ((\beta \Rightarrow \gamma) \Rightarrow (\alpha \Rightarrow \gamma))$,
(2) $(\gamma \Rightarrow \alpha) \Rightarrow ((\gamma \Rightarrow \beta) \Rightarrow (\gamma \Rightarrow (\alpha \sqcap \beta)))$,
(3) $(\alpha \sqcap \beta) \Rightarrow \alpha$ and $(\alpha \sqcap \beta) \Rightarrow \beta$,
(4) $\alpha \Rightarrow (\alpha \sqcup \beta)$ and $\beta \Rightarrow (\alpha \sqcup \beta)$,
(5) $(\alpha \Rightarrow \gamma) \Rightarrow ((\beta \Rightarrow \gamma) \Rightarrow ((\alpha \sqcup \beta) \Rightarrow \gamma))$,
(6) $(\alpha \& \beta) \Rightarrow (\beta \& \alpha)$,
(7) $(\alpha \& \beta) \Rightarrow \alpha$,
(8) $(\alpha \Rightarrow (\beta \Rightarrow \gamma)) \Rightarrow ((\alpha \& \beta) \Rightarrow \gamma)$,
(9) $((\alpha \& \beta) \Rightarrow \gamma) \Rightarrow (\alpha \Rightarrow (\beta \Rightarrow \gamma))$,
(10) $\overline{0} \Rightarrow \alpha$ and $\alpha \Rightarrow \overline{1}$.
The only rule of $\mathbf{FL_{ew}}$ is modus ponens:
$$\frac{\alpha, \alpha \Rightarrow \beta}{\beta}.$$

Other connectives in $\mathbf{FL_{ew}}$ can be defined as follows:
$$\neg \alpha = \alpha \Rightarrow \overline{0},$$
$$\alpha \leftrightarrow \beta \equiv (\alpha \Rightarrow \beta) \sqcap (\beta \Rightarrow \alpha).$$

Definition 2.2. *[5] Considering the following axiomatic extensions of* $\mathbf{FL_{ew}}$:
-*Intutitionistic logic* **IL** *is* $\mathbf{FL_{ew}}$ *plus the axiom*

$$(IDE) \quad \alpha \Rightarrow (\alpha \& \alpha).$$

-*Monoidal t-norm based logic* **MTL** *is* $\mathbf{FL_{ew}}$ *plus the axiom*

$$(PRE) \quad (\alpha \Rightarrow \beta) \sqcup (\beta \Rightarrow \alpha).$$

-*Basic fuzzy logic* **BL** *is* **MTL** *plus the axiom*

$$(DIV) \quad (\alpha \sqcap \beta) \Rightarrow (\alpha \& (\alpha \Rightarrow \beta)).$$

-*Gödel logic* **G** *is* **BL** *plus the axiom* (IDE).
-*Classical logic* **B** *is* $\mathbf{FL_{ew}}$ *plus the axiom*

$$(MID) \quad \neg \alpha \sqcup \alpha.$$

All of the mentioned logics are algebraizable, meaning that they are strongly complete with respect to their corresponding classes of algebras. Specifically, $\mathbf{FL_{ew}}$ is complete with respect to the variety $\mathbb{RL}$ of residuated lattices, **MTL** is complete with respect to the variety of $\mathbb{MTL}$ of MTL-algebras, which are equivalent to the variety of pre-linear residuated lattices, **IL** is complete with respect to the variety of $\mathbb{HA}$ of Heyting algebras, which are equivalent to idempotent residuated lattices, and **BL** is complete with respect to the variety of $\mathbb{BL}$ of BL-algebras, which are equivalent to divisibility MTL-algebras [4].

Definition 2.3. *[6] A residuated lattice is an algebra* $(L, \wedge, \vee, *, \rightarrow, 0, 1)$ *of type* $(2, 2, 2, 2, 0, 0)$ *such that* $(L, *, 1)$ *is a commutative monoid,* $(L, \wedge, \vee, 0, 1)$ *is a bounded lattice and the following residuation condition holds:*

$$x * y \leq z \text{ if and only if } x \leq y \rightarrow z,$$

where $\leq$ *is the order given by the lattice structure.*

3 Algebralization of substructural fuzzy epistemic logics

Inspired by Hájek's fuzzy epistemic logic **KD45(BL)** [15], we extend the language of substructural fuzzy logics $\mathbf{FL_{ew}}$ by introducing two logical connectives $\square$ and $\lozenge$. This extension is referred to as substructural fuzzy epistemic logic, denoted as $\mathbf{KD45(FL_{ew})}$. Moreover, we show that pseudo monadic residuated lattices serve as algebraic models of $\mathbf{KD45(FL_{ew})}$ and discuss some of their fundamental algebraic properties.

Definition 3.1. $\mathbf{KD45(FL_{ew})}$ *consists of all axiom schemes of* $\mathbf{FL_{ew}}$ *and the following axioms and rules:*

(KD1) $\Box 1$,
(KD2) $\neg \Diamond 0$,
(KD3) $\Box \alpha \Rightarrow \Diamond \alpha$,
(KD4) $\Box(\alpha \Rightarrow \Box \beta) \equiv \Diamond \alpha \Rightarrow \Box \beta$,
(KD5) $\Box(\Box \alpha \Rightarrow \beta) \equiv \Box \alpha \Rightarrow \Box \beta$
(KD6) $\Diamond \alpha \Rightarrow \Box \Diamond \alpha$,
(KD7) $\Box(\alpha \sqcap \beta) \equiv \Box \alpha \sqcap \Box \beta$,
(KD8) $\Diamond(\alpha \sqcup \beta) \equiv \Diamond \alpha \sqcup \Diamond \beta$,
(KD9) $\Diamond(\alpha \& \Diamond \beta) \equiv \Diamond \alpha \& \Diamond \beta$,
(KD10) $\Diamond(\Diamond \alpha \Rightarrow \Diamond \beta) \equiv \Diamond \alpha \Rightarrow \Diamond \beta$,
(KD11) $\Diamond(\Diamond \alpha \sqcap \Diamond \beta) \equiv \Diamond \alpha \sqcap \Diamond \beta$.

Remark 3.2. *It is worth noting that the usual semantics of fuzzy epistemic logic is a Kripke-style semantics. This is why in Hájek's famous book a fuzzy possibilistic semantics for a system of fuzzy epistemic logic is proposed. Unfortunately, it is not immediately evident how to derive an axiomatization of the underlying substructural fuzzy epistemic logic from this semantics, since both the axioms*

$$(K) \quad \Box(\alpha \Rightarrow \beta) \Rightarrow \Box \alpha \Rightarrow \Box \beta,$$
$$(C) \quad \Box \alpha \& \Box \beta \Rightarrow \Box(\alpha \& \beta),$$

are not valid. Therefore, we attack the problem in a novel way by introducing a possible algebraic semantics. This is achieved by extending residuated lattices (which is the most representative algebraic model of substructural fuzzy logic)) by two unary operators that model necessity and possibility.

In order to demonstrate that $\mathbf{KD45(FL_{ew})}$ is algebralize, we will utilize a general result derived from Abstract Algebraic Logicand start by showing that the logic is an implicative logic in the sense of Rasiowa. An implicative logic is a logic in which a connective $\Rightarrow$ exists in the logical language that satisfies the following conditions:

(R) $\vdash \alpha \Rightarrow \alpha$,
(MP) $\alpha, \alpha \Rightarrow \beta \vdash \beta$,
(T) $\alpha \Rightarrow \beta, \beta \Rightarrow \gamma \vdash \alpha \Rightarrow \gamma$,
(Cong) $\alpha \Rightarrow \beta, \beta \Rightarrow \alpha \vdash c(\gamma_1, \cdots, \gamma_i, \alpha, \cdots, \gamma_n) \Rightarrow c(\gamma_1, \cdots, \gamma_i, \beta, \cdots, \gamma_n)$,
(W) $\alpha \vdash \beta \Rightarrow \alpha$.

Most of these axioms hold trivially for $\mathbf{KD45(FL_{ew})}$ as they do for $\mathbf{FL_{ew}}$. Now, the consequential general result that can be applied is that $\mathbf{KD45(FL_{ew})}$ is algebraize and complete with respect to the variety of algebras known as pseudo monadic residuated lattices.

Definition 3.3. *An algebra* $(L, \wedge, \vee, *, \rightarrow, \Box, \Diamond, 0, 1)$ *of type* $(2,2,2,2,1,1,0,0)$ *is called a pseudo monadic residuated lattice if* $(L, \wedge, \vee, *, \rightarrow, 0, 1)$ *is a residuated lattice that also satisfies:*

(P1) $\Box 1 = 1$,
(P2) $\Diamond 0 = 0$,
(P3) $\Box a \rightarrow \Diamond a = 1$,
(P4) $\Box(a \rightarrow \Box b) = \Diamond a \rightarrow \Box b$,
(P5) $\Box(\Box a \rightarrow b) = \Box a \rightarrow \Box b$
(P6) $\Diamond a \rightarrow \Box \Diamond a = 1$,
(P7) $\Box(a \wedge b) = \Box a \wedge \Box b$,
(P8) $\Diamond(a \vee b) = \Diamond a \vee \Diamond b$,
(P9) $\Diamond(a * \Diamond b) = \Diamond a * \Diamond b$,
(P10) $\Diamond(\Diamond a \rightarrow \Diamond b) = \Diamond a \rightarrow \Diamond b$,
(P11) $\Diamond(\Diamond a \wedge \Diamond b) = \Diamond a \wedge \Diamond b$.

Pseudo monadic residuated lattices form a variety denoted by $\mathbb{PRL}$. To simplify notation, if L is a residuated lattice and we enrich it with a pseudomonadic structure, we denote the resulting algebra as $(L, \Box, \Diamond)$. It is evident that for each proper subvariety $\mathbb{V}$ of residuated lattices, the algebras in $\mathbb{PRL}$ whose RL-reducts are in $\mathbb{V}$ form a proper subvariety $\mathbb{PV}$ of $\mathbb{PBL}$. These algebras will be called pseudomonadic $\mathbb{V}$-algebras.

In their work [13], Rachůnek and Šalounová introduced the monadic residuated lattice as a structure $(L, \forall, \exists)$ satisfying the following identities:

(M1) $\forall a \rightarrow a = 1$,
(M2) $a \rightarrow \exists a = 1$,
(M3) $\forall(a \vee \exists b) = \forall a \vee \exists b$,
(M4) $\exists \forall a = a$,
(M5) $\exists(a * a) = \exists a * \exists a$,
(M6) $\exists(\exists a * \exists b) = \exists a * \exists b$,
(M7) $\forall(a \rightarrow \exists \exists b) = \exists a \rightarrow \exists b$,
(M8) $\forall(\exists a \rightarrow b) = \exists a \rightarrow \forall b$,
(M9) $\forall \forall a = \forall a$,

for any $a, b \in L$.

Moreover, we give an equivalent axioms of monadic residuated lattices.

Theorem 3.4. *Let L be a residuated lattice, $\forall$ and $\exists$ are two maps on L. Then $(L, \forall, \exists)$ is a monadic residuated lattice iff it satisfies the following conditions:*

(M1) $\forall a \rightarrow a = 1$,
(M3) $\forall(a \vee \exists b) = \forall a \vee \exists b$
(M5) $\exists(a * a) = \exists a * \exists a$,

(M10) $\forall(\forall a \to b) = \forall a \to \forall b$,
(M11) $\forall(a \to \forall b) = \exists a \to \forall b$.

It is important to note that the variety of monadic residuated lattices is the equivalent algebraic semantics of the monadic fragment of substructural predicate fuzzy logic, which is in turn equivalent to the fuzzy modal logic **S5(RL)**. It can be observed that every monadic residuated lattice is an pseudo monadic residuated lattice if taking $\forall = \Box$, $\exists = \Diamond$. Therefore, the variety $\mathbb{MRL}$ of monadic residuated lattices is a subvariety of $\mathbb{PRL}$. However, equations (M1), (M3) and (M5) are not valid in any pseudo monadic residuated lattice.

Example 3.5. *[14] Let L be the four-element MV-algebra over the universal $L = \{0, \frac{1}{3}, \frac{2}{3}, 1\}$ and the algebra $(L, \Box, \Diamond)$, where $\Box$ and $\Diamond$ are given by*

$$\Box x = \begin{cases} 0, & x = 0, \frac{1}{3}, \\ 1, & x = \frac{2}{3}, 1. \end{cases} \qquad \Diamond x = \begin{cases} 0, & x = 0, \frac{1}{3}, \\ 1, & x = \frac{2}{3}, 1. \end{cases}$$

It is easy to verified that $(L, \Box, \Diamond)$ satisfies the set of axioms of pseudo monadic residuated lattice, but the axiom (M1) is not valid for $a = \frac{2}{3}$, since

$$\Box \tfrac{2}{3} = 1 \not\leq \tfrac{2}{3}.$$

Example 3.6. *Let $L = \{0, a, b, c, d, 1\}$, where $0 \leq a, b$; $a \leq c, d$; $b \leq c$; $c, d \leq 1$. Defining operations $\to$ and $*$ as follows:*

$\to$	0	a	b	c	d	1
0	1	1	1	1	1	1
a	c	1	c	1	1	1
b	d	d	1	1	d	1
c	a	d	c	1	d	1
d	b	c	b	c	1	1
1	0	a	b	c	d	1

$*$	0	a	b	c	d	1
0	0	0	0	0	0	0
a	0	0	0	0	0	a
b	0	0	b	b	0	b
c	0	0	b	b	a	c
d	0	0	0	a	d	d
1	0	a	b	c	d	1

Then $(L, \wedge, \vee, *, \to, 0, 1)$ is a residuated lattice. Now, we define $\Box$ and $\Diamond$ as follows:

$$\Box x = \begin{cases} 1, & x = 1, \\ b, & x = b, c, \\ d, & x = d, \\ 0, & x = 0, a, \end{cases} \qquad \Diamond x = \begin{cases} 1, & x = 1, \\ b, & x = b, \\ d, & x = a, d, \\ 0, & x = 0. \end{cases}$$

It is easily to checked that $(L, \Box, \Diamond)$ *is a pseudo monadic residuated lattice. However, it does not satisfy axiom* $(M5)$, *since*

$$\Diamond(a * a) = \Diamond 0 = 0 \neq d = \Diamond a * \Diamond a.$$

Then we study some properties of pseudomonadic residuated lattice $(L, \Box, \Diamond)$.

Proposition 3.7. *Let* $(L, \Box, \Diamond)$ *be a pseudo monadic residuated lattice. Then we have: for any* $a, b \in L$,
 (P12) $\Box 0 = 0$,
 (P13) $\Diamond 1 = 1$,
 (P14) $\Box\Box a = \Box a$,
 (P15) $\Diamond\Box a = \Box a$,
 (P16) $\Diamond\Diamond a = \Diamond a$,
 (P17) $\Box\Diamond a = \Diamond a$,
 (P18) $\Diamond(\Diamond a \vee \Diamond b) = \Diamond a \vee \Diamond b$,
 (P19) $\Diamond(\Diamond a * \Diamond b) = \Diamond a * \Diamond b$,
 (P20) $\Box(\Diamond a \to b) = \Diamond a \to \Box b$,
 (P21) $\Box \neg a = \neg \Diamond a$,
 (P22) $\Box(\Box a \to a) = 1$,
 (P23) $\Box(a \to \Diamond a) = 1$,
 (P24) *If* $a \to b = 1$, *then* $\Box a \to \Box b = 1$,
 (P25) *If* $a \to b = 1$, *then* $\Diamond a \to \Diamond b = 1$.

Proof. (P12) We can derive from (P2) and (P3) the inequality $\Box 0 \leq \Diamond 0 = 0$, which implies $\Box 0 = 0$.
 (P13) From (P1) and (P3), we have $1 = \Box 1 \leq \Diamond 1$, therefore $\Diamond 1 = 1$.
 (P14) By applying (P3), we can deduce $\Box\Box a \to \Diamond\Box a = 1$, i.e., $\Box\Box a \leq \Diamond\Box a$. Moreover, using (P1) and (P4), we have

$$1 = \Box(\Box a \to \Box a) = \Diamond\Box a \to \Box a,$$

that is $\Diamond\Box a \leq \Box a$. This leads to $\Box\Box a \leq \Box a$.
 On the other hand, by (P1) and (P5), we have

$$1 = \Box(\Box a \to \Box a) = \Box a \to \Box\Box a,$$

which implies $\Box a \leq \Box\Box a$, and consequently, $\Box\Box a = \Box a$.
 (P15) It is immediate from the previous proof (P14).
 (P16) From (P9), we get

$$\Diamond\Diamond a = \Diamond(1 * \Diamond a) = \Diamond 1 * \Diamond a = \Diamond a,$$

that is, $\Diamond\Diamond a = \Diamond a$.

(P17) Using (P6), we have $\Diamond a \leq \Box\Diamond a$. For the other direction, considering (P3) and (P16), we have

$$\Box\Diamond a \leq \Diamond\Diamond a = \Diamond a.$$

Hence $\Box\Diamond a = \Diamond a$.

(P18) This is a direct consequence of (P8) and (P16).

(P19) It is an immediate consequence of (P9) and (P16).

(P20) By applying (P17) and (P5), we get

$$\Box(\Diamond a \to b) = \Box(\Box\Diamond a \to b) = \Box\Diamond a \to \Box b = \Diamond a \to \Box b.$$

(P21) By combining (P12) and (P4), we obtain

$$\Box\neg a = \Box(a \to 0) = \Box(a \to \Box 0) = \Diamond a \to \Box 0 = \Diamond a \to 0 = \neg\Diamond a.$$

(P22) Taking into account (P5), we have $\Box(\Box a \to a) = \Box a \to \Box a = 1$.

(P23) From (P17) and (P4), we have

$$\Box(a \to \Diamond a) = \Box(a \to \Box\Diamond a) = \Diamond a \to \Box\Diamond a = \Diamond a \to \Diamond a = 1,$$

that is, $\Box(a \to \Diamond a) = 1$.

(P24) It is an immediate consequence of (P7).

(P25) It is an immediate consequence of (P8). $\square$

Proposition 3.8. *Let $(L, \Box, \Diamond)$ be a pseudo monadic residuated lattice. Then $\Box L = \Diamond L$, and $\Diamond L$ is a subalgebra of L.*

Proof. By using (P15) and (P17), we can conclude that

$$\Box L = \{\Box a : a \in L\} = \{\Diamond a : a \in L\} = \Diamond L.$$

On the other hand, from (P2), (P13), (P16), (P18), (P19), (P10) and (P11), we obtain that $\Diamond L$ is a subalgebra of L. $\square$

Remark 3.9. *It is noted that (P14) and (P16) imply that $\Box$ and $\Diamond$ are idempotent operations, which means they are equal when restricted to the subalgebra $\Diamond L$. Furthermore, (P24) and (P25) demonstrate that both operators are monotonic.*

A nonempty subset F of a residuated lattice L is called a *filter* if it satisfies: (1) $1 \in F$; (2) $a \in F$ and $a \to b \in F$ imply $b \in F$. A filter F of L is called a *proper filter* if $F \neq L$. Every filter F of a residuated lattice L determines a congruence $\equiv_F$ given by

$$a \equiv_F b \text{ if and only if } a \to b \in F \text{ and } b \to a \in F.$$

Moreover, the map $F \mapsto \equiv_F$ is an order isomorphism between the lattice of filters of a residuated lattice L and the lattice of congruences of L. Now we will generalize the notion of filters for our new structures.

Definition 3.10. *A subset F of a pseudo monadic residuated lattice $(L, \Box, \Diamond)$ is a pseudo monadic filter if F is a filter and if $a \to b \in F$, then $\Box a \to \Box b \in F$ and $\Diamond a \to \Diamond b \in F$.*

Theorem 3.11. *Let F be a pseudo monadic filter of a pseudo monadic residuated lattice $(L, \Box, \Diamond)$. Then, the binary relation $\equiv_F$ on L defined by $a \equiv_F b$ if and only if $a \to b \in F$ and $b \to a \in F$ is a congruence relation. Moreover, $F = \{a \in L : a \equiv_F 1\}$. Conversely, if $\equiv$ is a congruence on L, then $F_\equiv = \{a \in L : a \equiv 1\}$ is a pseudo monadic filter, and $a \equiv b$ if and only if $a \to b = 1$ and $b \to a = 1$. Therefore, the correspondence $F \mapsto \equiv_F$ is a bijection from the set of pseudo monadic filters of $(L, \Box, \Diamond)$ onto the set of congruences on $(L, \Box, \Diamond)$.*

Proof. The fact that the congruence $\equiv_F$ of a residuated lattice L is also a congruence of the pseudo monadic residuated lattice $(L, \Box, \Diamond)$ follows immediately from the definition of pseudo monadic filters. We will check that $F = \{a \in L : a \equiv_F 1\}$. In detail, $a \to 1 = 1 \in F$ and if $a \in F$, since $a = 1 \to a$, we have $1 \to a \in F$. Hence, $a \equiv_F 1$. On the other hand, if we consider $a \in \{a \in L : a \equiv_F 1\}$, it is immediate that $a = 1 \to a \in F$. $\square$

4 Subvarieties of pseudo monadic residuated lattices

In this section, we will see that pseudo monadic residuated lattices generalize three well studied classes of algebras. In particular, the subclass that we are interested in is the algebraic counterpart of modal fuzzy logic **KD45**. First, given a pseudo monadic residuated lattice $(L, \Box, \Diamond)$, we will show that

(1) if the reduct L is a BL-algebra, then the algebra $(L, \Box, \Diamond)$ is a pseudo monadic BL-algebra,

(2) if the reduct L is a Gödel algebra, then the algebra $(L, \Box, \Diamond)$ is a Bi-modal Gödel algebra,

(3) if the reduct L is a Boolean algebra, then the algebra $(L, \Diamond)$ is a pseudo monadic algebra.

In [15], the authors introduced the class of pseudo monadic BL-algebras that they serve as algebraic models of Hájek's fuzzy modal logic.

Definition 4.1. *[15] An algebra $(L, \wedge, \vee, *, \to, \Box, \Diamond, 0, 1)$ of type $(2, 2, 2, 2, 1, 1, 0, 0)$ is called a pseudo monadic BL-algebra if $(L, \wedge, \vee, *, \to, 0, 1)$ is a BL-algebra that satisfies:*

(PBL1) $\Box 1 = 1$,
(PBL2) $\Diamond 0 = 0$,
(PBL3) $\Box a \to \Diamond a = 1$,
(PBL4) $\Box(a \to \Box b) = \Diamond a \to \Box b$,
(PBL5) $\Box(\Box a \to b) = \Box a \to \Box b$,
(PBL6) $\Diamond a \to \Box \Diamond a = 1$,
(PBL7) $\Box(a \vee b) = \Box a \vee \Box b$,
(PBL8) $\Box(a \wedge b) = \Box a \wedge \Box b$.
(PBL9) $\Diamond(a * \Diamond b) = \Diamond a * \Diamond b$.

The class of pseudo monadic BL-algebras forms a variety which is denoted by $\mathbb{PBL}$.

Theorem 4.2. *Let L be a BL-algebra. Then $(L, \Box, \Diamond)$ is a pseudo monadic BL-algebra iff $(L, \Box, \Diamond)$ is a pseudo monadic residuated lattice.*

Proof. Assume first that $(L, \Box, \Diamond)$ is a pseudo monadic residuated lattice, we recall that since L is a BL-algebra, it satisfies the conditions:

$$(\text{DIV}) \quad a \wedge b = a * (a \to b),$$
$$(\text{PRE}) \quad (a \to b) \vee (b \to a) = 1.$$

(P11) By (P3) and (P20), we have that

$$\Diamond a \to \Diamond b = \Diamond a \to \Box \Diamond b = \Box(\Diamond a \to \Diamond b) \leq \Diamond(\Diamond a \to \Diamond b),$$

which implies $\Diamond a \to \Diamond b \leq \Diamond(\Diamond a \to \Diamond b)$. Moreover, by applying (P17),(P4) and(P1), we can deduce

$$\Diamond(\Diamond a \wedge \Diamond b) \to \Diamond b = \Diamond(\Diamond a \wedge \Diamond b) \to \Box \Diamond b$$
$$= \Box((\Diamond a \wedge \Diamond b) \to \Box \Diamond b)$$
$$= \Box((\Diamond a \wedge \Diamond b)) \to \Diamond$$
$$= \Box 1$$
$$= 1,$$

and consequently, $\Diamond(\Diamond a \wedge \Diamond b) \leq \Diamond b$. By the condition (DIV) and (P9), we have

$$\Diamond(\Diamond a * (\Diamond a \to \Diamond b)) \leq \Diamond b, \; \Diamond(\Diamond a \to \Diamond b) * \Diamond a \leq \Diamond b,$$

further by residuation,

$$\Diamond(\Diamond a \to \Diamond b) \leq \Diamond a \to \Diamond b.$$

(P12) By the condition (DIV), we have

$$\Diamond(\Diamond a \wedge \Diamond b) = \Diamond(\Diamond a * (\Diamond a \to \Diamond b)).$$

Then by (P11), the right side of the last identity is equivalent to
$$\Diamond(\Diamond a * \Diamond(\Diamond a \to \Diamond b)),$$
further by (P19), which is equivalent to
$$\Diamond a * \Diamond(\Diamond a \to \Diamond b) = \Diamond a * (\Diamond a \to \Diamond b) = \Diamond a \wedge \Diamond b.$$

Then follows from Definition 4.1 that $(L, \Box, \Diamond)$ is a pseudo monadic BL-algebra. Conversely is trivial. $\Box$

Bezhanishvili introduced the class of pseudo monadic algebras as natural generalizations of monadic algebras and showed that they serves as algebraic version of **KD45** over classical logic [16].

Definition 4.3. *[16] An algebra $(L, \Diamond)$ is said to be a pseudo monadic algebra if L is a Boolean algebra and $\Diamond$ is a unary operator on L satisfying the following identities:*
  (PB1) $\Diamond 0 = 0$,
  (PB2) $\Diamond(a \vee b) = \Diamond a \vee \Diamond b$,
  (PB3) $\Diamond(\Diamond a \wedge b) = \Diamond a \wedge \Diamond b$,
  (PB4) $\neg \Diamond a \leq \Diamond \neg a$,
for any $a, b \in L$.

The class of pseudo monadic algebras forms a variety denoted by $\mathbb{PMA}$, which is a proper extension of the variety of monadic algebras. In this case, we use $\Box$ as an abbreviation of the operator $\neg \Diamond \neg$ since they are dual.

Theorem 4.4. *Let L be a Boolean algebra. Then $(L, \Box, \Diamond)$ is a pseudo monadic residuated lattice if and only if $(L, \Diamond)$ is a pseudo monadic algebra.*

Proof. Notice that Busaniche et al. have proven that if L is a Boolean algebra, then $(L, \Box, \Diamond)$ is a pseudo monadic BL-algebra iff $(L, \Diamond)$ is a pseudo monadic algebra. Further by Theorem 4.2, the result of this theorem can be proved. $\Box$

Gödel algebras can be characterized as the subvariety of residuated lattices determined by the equation
$$a * a = a, \quad a \wedge b = a * (a \to b).$$

Caicedo and Rodriguez showed that the set of valid formulas in the subclass of serial, transitive and Euclidean GK-frames (Gödel Kripke frames) can be axiomatized by adding some additional axioms and a rule to those of Gödel fuzzy logic G. The logic obtained is denoted **KD45(G)** and has the variety of Bimodal Gödel algebras as its algebraic semantics. Notice that Busaniche et al. proved that the class of pseudo monadic Gödel algebras and the Bimodal Gödel algebras coincide [11].

Definition 4.5. *[11] An algebra* $(L, \wedge, \vee, \rightarrow, \Box, \Diamond, 0, 1)$ *of type* $(2,2,2,1,1,0,0)$ *is called a pseudo monadic Gödel algebra if* $(L, \wedge, \vee, \rightarrow, 0, 1)$ *is a Gödel algebra that satisfies the following conditions:*

(PG1) $\Box(a * b) = \Box a * \Box b$,
(PG2) $\Box 1 = 1$,
(PG3) $\Diamond a \rightarrow \Box b \leq \Box(a \rightarrow b)$,
(PG4) $\Diamond(a \vee b) = \Diamond a \vee \Diamond b$,
(PG5) $\Diamond 0 = 0$,
(PG6) $\Diamond(a \rightarrow b) \leq \Box a \rightarrow \Diamond b$,
(PG7) $\Box a \leq \Diamond a$,
(PG8) $\Box a \leq \Box\Box a$, $\Diamond a \leq \Diamond\Diamond a$,
(PG9) $\Diamond a \leq \Box \Diamond a$, $\Diamond \Box a \leq \Box a$.

Theorem 4.6. *Let L be a Gödel algebra. Then $(L, \Box, \Diamond)$ is a pseudo monadic residuated lattice if and only if $(L, \Diamond)$ is a pseudo monadic Gödel algebra.*

Proof. The results can be directly proved by Theorem 4.2 and Theorem 6 in [15]. □

5 Constructions of pseudo monadic residuated lattices

In this section, we offer a characterization of pseudomonadic residuated lattices as pairs of residuated lattices (L, B), where B is a special case of a relatively complete subalgebra of L called c-relatively complete, and give a necessary and sufficient condition for a subalgebra to be c-relatively complete. This results will become important to establish a connection with possibilistic RL-frames, which is similar to possibilistic BL-frames in [15].

Definition 5.1. *Let $(L, \Box, \Diamond)$ be a pseudo monadic residuated lattice, if the set*

$$\{a \in L : \Box a = 1\} \subsetneq L$$

has a least element c, then c will be called a focal element of L.

Example 5.2. *Let $(L, \Box, \Diamond)$ be a pseudo monadic residuated lattice in Example 3.5. Then*

$$\min\{a \in L | \Box a = 1\} = 1 \subsetneq L,$$

which implies that the focal element exists and it is 1.

Remark 5.3. *In fact, if $(L, \Box, \Diamond)$ is a pseudo monadic residuated lattice and L is finite, the focal element exists. However, this is not the case for every pseudo monadic residuated lattice, as demonstrated by the following example. Let L be the standard MV-algebra with the operators $\Box$ and $\Diamond$ defined by*

$$\Box x = \Diamond x = \begin{cases} 1, & x = (0,1], \\ 0, & x = 0. \end{cases}$$

Then resultant structure is a pseudo monadic residuated lattice such that the set

$$\{a \in L | \Box a = 1\} = (0,1]$$

has no least element.

Proposition 5.4. *Let c be a focal element of a pseudo monadic residuated lattice $(L, \Box, \Diamond)$. Then c satisfies*

$$c = \min\{(\Box a \to a) \wedge (a \to \Diamond a), a \in L\}. \qquad (C1)$$

Proof. Let x be a element of the form

$$x = (\Box a \to a) \wedge (a \to \Diamond a)$$

for some $a \in L$. Then it follows from (P5) and (P6) that

$$\begin{aligned} \Box x &= \Box((\Box a \to a) \wedge (a \to \Diamond a)) \\ &= \Box(\Box a \to a) \wedge \Box(a \to \Diamond a) \\ &= (\Box a \to \Box a) \wedge (\Diamond a \to \Diamond a) \\ &= 1, \end{aligned}$$

which implies $x \in \{x \in L : \Box x = 1\}$. Hence by the definition of the focal element, we have $c \leq x$. On the other hand, since $\Box c = \Diamond c = 1$, we can take

$$c = (\Box c \to c) \wedge (c \to \Diamond c).$$

Therefore, c is the least element of $(\Box a \to a) \wedge (a \to \Diamond a)$. $\square$

Definition 5.5. *A pseudo monadic residuated lattice $(L, \Box, \Diamond)$ with focal element c will be called a c-pseudo monadic residuated lattice.*

Example 5.6. *Let $(L, \Box, \Diamond)$ be a pseudo monadic residuated lattice such that L is finite. Then $(L, \Box, \Diamond)$ is a c-pseudo monadic residuated lattice. For example, pseudo monadic residuated lattices in Examples 3.5 and 3.6 are c-pseudo monadic residuated lattices.*

Regarding the class of c-pseudo monadic residuated lattices, the focal element plays an important role, since it allows us to recover the unary operators $\Box$ and $\Diamond$, as shown in the following theorem.

Theorem 5.7. *Let L be a c-pseudo monadic residuated lattice and B be the subalgebra given by Proposition 3.8. Then*

$$\Box a = \max\{b \in B : b \leq c \to a\},$$
$$\Diamond a = \min\{b \in B : c * a \leq b\}.$$

Proof. Since $B = \Box L = \Diamond L$ and c satisfies the condition $(C1)$, we get

$$c \leq (\Box a \to a) \wedge (a \to \Diamond a) \leq \Box a \to a$$

for all $a \in L$. By residuation, $\Box a \leq c \to a$. Now we assume that there is $b \in B$ such that $b \leq c \to a$. Then there exists $x \in L$ such that $b = \Box x$, and so $\Box x \leq c \to a$, or equivalently, $c \leq \Box x \to a$. By (P13), we have

$$\Box c \leq \Box x \to \Box a.$$

Since $\Box c = 1$, it follows that $b = \Box x \leq \Box a$. Therefore, $\Box a = \max\{b \in B : b \leq c \to a\}$.

Arguing as above,

$$c \leq (\Box a \to a) \wedge (a \to \Diamond a) \leq a \to \Diamond a$$

for all $a \in L$. Thus $c * a \leq \Diamond a$. Suppose there exists $b \in B$ such that $c * a \leq b$ with $b = \Diamond y$ for some $y \in L$. By residuation, $c \leq a \to \Diamond y$. By (P13) and (P6), we obtain $\Box c \leq \Diamond a \to \Diamond y$, and then

$$\Diamond a \leq \Box c \to \Diamond y = 1 \to \Diamond y = \Diamond y = b.$$

We can conclude that $\Diamond a = \min\{b \in B : c * a \leq b\}$. □

According to Proposition 3.8, if $(L, \Box, \Diamond)$ is a pseudo monadic residuated lattice, then $\Diamond L = \Box L$ is a subalgebra of L. We are going to show under which conditions a pseudo monadic residuated lattice can be defined from a residuated lattic L and one of its subalgebras B.

Definition 5.8. *Let L be a residuated lattice, B be a subalgebra of L and $c \in L$. Then the pair (B, c) is a c-relative complete subalgebra, if the following conditions hold:*

(e1) For any $a \in L$, the subset

$$\{b \in B : b \leq c \to a\}$$

has a greatest element, and the subset

$$\{b \in B : c * a \leq b\}$$

has a least element.

(e2) $\{a \in L : c^2 \leq a\} \cap B = \{1\}$.

Theorem 5.9. *Let L be a residuated lattice and (B,c) be a c-relative complete subalgebra. If we define the operations on L:*

$$\Box a = \max\{b \in B : b \leq c \to a\}, \qquad (\Box 1)$$
$$\Diamond a = \min\{b \in B : c * a \leq b\}. \qquad (\Diamond 1)$$

Then $(L, \Diamond, \Box)$ is a c-pseudo monadic residuated lattice such that $\Box L = \Diamond L = B$. Conversely, if L is a c-pseudo monadic residuated lattice, then $(\Box L, c)$ is a c-relative complete subalgebra of L.

Proof. Clearly condition (e1) guarantees the existence of $\Box a$ and $\Diamond a$ for every $a \in L$. It remains to show that $(L, \Box, \Diamond)$ satisfies Definition 3.1. Let $a, b \in L$,

(P1) Because B is a subalgebra, so it is clear that

$$\Box 1 = \max\{b \in B : b \leq c \to 1\} = 1,$$

that is, $\Box 1 = 1$.

(P2) Similarly, we can get

$$\Diamond 0 = \min\{b \in B : c * 0 \leq b\} = 0,$$

that is, $\Diamond 0 = 0$.

(P3) By definition, we have $\Box a \leq c \to a$. Then, $c * \Box a \leq a$ and $c^2 * \Box a \leq c * a$. Since $c * a \leq \Diamond a$, we can get $c^2 * \Box a \leq \Diamond a$. By residuation, $c^2 \leq \Box a \to \Diamond a$. Besides, from $(\Diamond 1)$ and $(e2)$, we have $\Diamond c = 1$ and

$$\Diamond c = \min\{b \in B : c^2 \leq b\} = 1.$$

Since $\Box a \to \Diamond a \in B$, hence

$$1 = \Diamond c \leq \Box a \to \Diamond a.$$

That is, $\Box a \to \Diamond a = 1$.

(P4) On the one hand, from $a * c \leq \Diamond a$, we can get

$$\Diamond a \to \Box b \leq (a * c) \to \Box b = c \to (a \to \Box b).$$

Since $\Box a \to \Diamond a \in B$, we obtain

$$\Diamond a \to \Box b \leq \Box(a \to \Box b).$$

On the other hand, we know that

$$\Box(a \to \Box b) \leq c \to (a \to \Box b) = (a * c) \to \Box b.$$

Hence, by residuation,
$$a * c \leq \Box(a \to \Box b) \to \Box b.$$
Taking into account ($\Diamond 1$), we have
$$\Diamond a \leq \Box(a \to \Box b) \to \Box b,$$
so $\Box(a \to \Box b) \leq \Diamond a \to \Box b$.

(P5) By definition,
$$\Box(\Box a \to b) \leq c \to (\Box a \to b) = \Box a \to (c \to b),$$
that is, $\Box(\Box a \to b) * \Box a \leq c \to b$. Taking into account ($\Box 1$), we obtain
$$\Box(\Box a \to b) \leq \Box a \to \Box b.$$
Moreover, $\Box b \leq c \to b$ implies
$$\Box a \to \Box b \leq \Box a \to (c \to b) = c \to (\Box a \to b),$$
from where, by definition of $\Box(\Box a \to b)$, we get $\Box a \to \Box b \leq \Box(\Box a \to b)$.

(P6) We know that $\Diamond a \leq c \to \Diamond a$ and $\Diamond a \in B$. Therefore, $\Diamond a \leq \Box \Diamond a$, that is $\Diamond a \to \Box \Diamond a = 1$.

(P7) Note that
$$\Box(a \wedge b) \leq c \to (a \wedge b) = (c \to a) \wedge (c \to b).$$
Then $\Box(a \wedge b) \leq c \to a = \Box a$ and $\Box(a \wedge b) \leq c \to b = \Box b$. Hence $\Box(a \wedge b) \leq \Box a \wedge \Box b$.
On the other hand, $\Box a \wedge \Box b \leq \Box a \leq c \to a$ and $\Box a \wedge \Box b \leq \Box b \leq c \to b$. Thus,
$$\Box a \wedge \Box b \leq (c \to a) \wedge (c \to b) = c \to (a \wedge b).$$
Since $\Box(a \wedge b) = \max\{a \in B : a \leq c \to (a \wedge b)\}$, we obtain $\Box a \wedge \Box b \leq \Box(a \wedge b)$.

(P8) The proof is analogous to the (P7).

(P9) By definition $c * (a * \Diamond b) \leq \Diamond(a * \Diamond b)$, then by residuation,
$$c * a \leq \Diamond b \to \Diamond(a * \Diamond b).$$
So $\Diamond a \leq \Diamond b \to \Diamond(a * \Diamond b)$, or equivalently $\Diamond a * \Diamond b \leq \Diamond(a * \Diamond b)$. Since $a * c \leq \Diamond a$, we have
$$(a * c) * \Diamond b \leq \Diamond a * \Diamond b.$$
Therefore, $\Diamond(a * \Diamond b) \leq \Diamond a * \Diamond b$.

(P10) Since $c * (\Diamond a \to \Diamond b) \leq \Diamond(\Diamond a \to \Diamond b)$. Then by residuation, we have
$$\Diamond a \to \Diamond b \leq c \to \Diamond(\Diamond a \to \Diamond b).$$

According to ($\Box 1$), we can get $\Diamond a \to \Diamond b \leq \Box\Diamond(\Diamond a \to \Diamond b)$. Hence $\Diamond a \to \Diamond b \leq \Diamond(\Diamond a \to \Diamond b)$. On the other hand, we have

$$\Box(\Diamond a \to \Diamond b) \leq c \to (\Diamond a \to \Diamond b).$$

So $\Box(\Box\Diamond a \to \Diamond b) \leq c \to (\Diamond a \to \Diamond b)$, or equivalently, $\Box\Diamond a \to \Diamond b \leq c \to (\Diamond a \to \Diamond b)$. Hence

$$\Diamond a \to \Diamond b \leq c \to (\Diamond a \to \Diamond b).$$

By residuation, we have

$$c * (\Diamond a \to \Diamond b) \leq \Diamond a \to \Diamond b.$$

Therefore, $\Diamond(\Diamond a \to \Diamond b) \leq \Diamond a \to \Diamond b$.

(P11) Since $c * (\Diamond a \wedge \Diamond b) \leq \Diamond(\Diamond a \wedge \Diamond b)$, by residuation, we can get

$$\Diamond a \wedge \Diamond b \leq c \to \Diamond(\Diamond a \wedge \Diamond b).$$

Then $\Diamond a \wedge \Diamond b \leq \Box\Diamond(\Diamond a \wedge \Diamond b)$, that is, $\Diamond a \wedge \Diamond b \leq \Diamond(\Diamond a \wedge \Diamond b)$. Moreover,

$$c * (\Diamond a \wedge \Diamond b) \leq \Diamond a \wedge \Diamond b.$$

So $\Diamond(\Diamond a \wedge \Diamond b) = \Diamond a \wedge \Diamond b$.

Thus, $\langle L, \Box, \Diamond \rangle$ is a pseudo monadic residuated lattice.

We verify that c is the focal element of $\langle L, \Box, \Diamond \rangle$. To that aim, let a be an element of $\{a \in L : \Box a = 1\}$. Then, from ($\Box 1$), we get $1 = \Box a \leq c \to a$ and $c \leq a$. Besides,

$$\leq \Box c = \max\{b \in B : b \leq c \to c\} = 1,$$

and so $c = \min\{a \in L : \Box a = 1\}$.

Now let us see that $\Box L = \Diamond L = B$. By the previous and Theorem 3.4, the first equality is satisfied. On the other hand, it is clear that $\Diamond L \subseteq B$. Furthermore, for all $b \in B$, $c * b \leq b$, whereby $\Diamond b \leq b$. Besides $c^2 * b \leq c * b \leq \Diamond b$. Then by residuation, $c^2 \leq b \to \Diamond b$, but $b, \Diamond b \in B$ and B is subalgebra, it follows that $b \to \Diamond b \in B$ and is greater than c^2. Thus, $1 = \Diamond c \leq b \to \Diamond b$. Consequently, $b \leq \Diamond b$, and hence $B \subseteq \Diamond L$.

Conversely, let $\langle L, \Box, \Diamond \rangle$ be a c-pseudomonadic residuated lattice. From Theorem 3.4, we know that $\Diamond L$ is a subalgebra of L. Let us now show that conditions (e1) and (e2) hold.

(e1) By Theorem 4.4, $\Box a = \max\{b \in \Box L : b \leq c \to a\}$ and $\Diamond a = \min\{b \in \Box L : c * a \leq b\}$.

(e2) $\{a \in L : c^2 \leq a\} \cap \Box L = \{a \in \Box L : c^2 \leq a\}$. By Theorem 4.5, the set $\{a \in \Box L : c^2 \leq a\}$ has a least element and is $\Diamond c$, and so $\{a \in \Box L : c^2 \leq a\} = \{1\}$. $\Box$

6 Conclusions

The motivation behind our paper is to present an algebraic characterization of a fuzzy epistemic logic system that extends the classical **KD45**, and it is based on substructural fuzzy epistemic logics. To achieve this goal, we have introduced pseudo monadic residuated lattices, as residuated lattices with two unary operators that behave generalizing the modal operators of **KD45**. We have studied some of their logical and algebraic properties, and we have shown their relationship with monadic residuated lattices, which turn to be the algebraic counterpart of the fuzzy version of **S5**. The results of Section 4 suggest that our definition of pseudo monadic residuated lattices is on firm ground: We have shown that pseudo monadic residuated lattices whose RL-reduct are Boolean algebras coincide with the algebraic correspondent of classical **KD45** (Pseudomonadic algebras) and that the ones whose RL-reduct is a Gödel algebra are equivalent to serial transitive and Euclidean Bimodal Gödel algebras, the algebraic correspondent to the Gödel generalization of **KD45**. To close the ideas of the paper, after investigating c-pseudo monadic residuated lattices and complex pseudo monadic residuated lattices, we give a construction of c-pseudo monadic residuated lattices.

References

[1] C. C. Chang. Algebraic analysis of many-valued logics. *Transactions of the American Mathematical Society*, 88(2): 467–490, 1958.

[2] P. Hájek. *Metamathematics of Fuzzy Logic*. Kluwer Academic Publishers, Dordrecht, 1998.

[3] R. Cignoli, F. Esteva, L. Godo and A. Torrens. Basic Fuzzy Logic is the logic of continuous t-norms and their residua. *Soft Computing*, 4: 106–112, 2000.

[4] F. Esteva and L. Godo. Monoidal t-norm based logic: towards a logic for left-continuous t-norms. *Fuzzy Sets and Systems*, 124(3): 271–288, 2001.

[5] S. Jenei and F. Montagan. A proof of standard completeness for Esteva and Godo's logic. *Studia Logica*, 70: 183-192, 2002.

[6] M. Ward and P. R. Dilworth. Residuated lattices. *Transactions of the American Mathematical Society*, 45: 335-354, 1939.

[7] Y. Q. Zhu and Y. Xu. On filter theory of residuated lattices. *Information Sciences*, 180(19): 3614-3632, 2010.

[8] K. J. J. Hintikka. Knowledge and belief: an introduction to the logic of the two notions. *Studia Logica*, 16: 119–122, 1962.

[9] M. C. Fitting. Many valued modal logics. *Fundamenta Informaticae*, 15(3-4): 254–325, 1991.

[10] M. C. Fitting. Many valued modal logics II. *Fundamenta Informaticae*, 17(1-2): 55–73, 1992.

[11] X. Caicedo and R. O. Rodriguez. Bi-modal Gödel logic over [0,1]-valued Kripke frames. *Journal of Logic and Computation*, 25(1): 37–55, 2015.

[12] F. Bou, F. Esteva, L. Godo and R. O. Rodriguez. On the minimum many valued modal logic over a finite residuated lattice. *Journal of Logic and Computation*, 21(5): 739–790, 2011.

[13] J. Rachůnek and D. Šalounová. Monadic bounded residuated lattices. *Order*, 30(1): 195–210, 2013.

[14] J. T. Wang, P. F. He and Y. H. She. Monadic NM-algebras. *Logic Journal of the IGPL*, 27(6): 812–835, 2019.

[15] M. Busaniche, P. Cordero and R. O. Rodriguez. Pseudomonadic BL-algebras: an algebraic approach to possibilistic BL-logic. *Soft Computing*, 23: 2199–2212, 2019.

[16] N. Bezhanishvili. Pseudomonadic algebras as algebraic models of doxastic modal logic. *Mathematical Logic Quarterly*, 48(4): 624–636, 2002.

Algebras of Similarity Monadic Fuzzy Predicate Logic

Xuesong Fu
School of Science, Xi'an Shiyou University, China
School of Mathematics and Computing Science, Xiangtan University, China
fuxs@smail.xtu.edu.cn

Xiaoyan Liu*
School of Science, Xi'an Shiyou University, China
feng20014100163.com

Zhiqin Zhao
School of Science, Xi'an Shiyou University, China
zhqinzhao2014@163.com

Abstract

In this paper, we introduce the notion of similarity monadic MTL-algebras and study some of their related basic algebraic properties. Then we introduce and investigate similarity monadic filters of similarity monadic MTL-algebras. In particular, by means of similarity monadic filters, we give some characterizations of representable similarity monadic MTL-algebras. Finally, we introduce the logic of similarity monadic MTL-algebras and prove the completeness of them.

keyword: Mathematical fuzzy logic, monadic fuzzy logical algebra, similarity, representation, completeness

---

The authors would like to express their grateful to anonymous reviewers for their comments and effort.

*Corresponding author.

1 Introduction

It is well-known that certain information processing approaches, especially inferences based on certain information, are based on the classical logic. Naturally, it is necessary to establish some rational logic systems as a logical foundation for uncertain information processing. For this reason, kinds of non-classical logic systems have been proposed and researched [6, 19, 11]. As semantic systems for non-classical logic systems, various logical algebras have been introduced and investigated, such as Hájek [19] presented a logic intended to be the basic fuzzy logic **BL** and gave an algebraic semantics for them introducing the variety of BL-algebras. Indeed, **BL** is a general framework of fuzzy logic for capturing the tautologies of continuous t-norm and their residua. However, it is proved that the sufficient and necessary condition for a t-norm to have a residuated implication is the left-continuity, hence it makes sense to consider fuzzy logics based not on continuous t-norm but on left continuous t-norms. Based on the above consideration, Esteva and Godo [11] proposed a new logic, called monoidal t-norm-based logic **MTL**, as the basic fuzzy logic in this more general sense, and gave an algebraic semantics for **MTL** introducing the variety of MTL-algebras. Afterwards, Jenei and Montagna [23] proved that **MTL** is indeed the logic of all left-continuous t-norms and their residua. Thus MTL-algebras are the most fundamental residuated structures contain all algebras induced by left continuous t-norms and their residua.

Monadic Boolean algebra $(L, \exists)$, in the sense of Halmos [21], is a Boolean algebra equipped with a closure operator $\exists$, which abstracts algebraic properties of the standard existential quantifier "for some". The name "monadic" comes from the connection with predicate logics for languages having one placed predicates and a single quantifier. After that, monadic MV-algebras, the algebraic counterpart of monadic Łukasiewicz logic, were introduced and studied in [28, 25]. Monadic BL-algebras, monadic residuated lattices, monadic residuated ℓ-monoids, monadic bounded hoops, monadic NM-algebras, monadic pseudo BCI-algebras and monadic pseudo equality algebras were introduced and investigated in [4, 8, 18, 26, 27, 32, 33, 42, 41, 17]. As for this topic, Wang recently has made some very interesting and meaningful explorations on the representations of monadic algebras and the completeness of their corresponding logics in [34, 35, 36, 37, 38, 39, 40]. And in particular, He has used the Kalman functor to relate the category of weak monadic residuated distributive lattices and the category of monadic c-differential residuated distributive lattices in [37], providing a new algebraic proof of completeness for monadic fuzzy predicate logic **MMTL∀** in [38], studying some deeper algebraic results of monadic BL-algebras in [34, 35, 36] and creatively study the algebraic semantics of similarity in monadic substructural predicate logics in [39].

The concept of similarity was introduced by Zadeh [43] to extend to the fuzzy framework the notion of equivalence relations. Since then similarity has been used for a wide range of applications, for example, in clustering, fuzzy control, fuzzy logic programming and in all the contexts in which there is the necessity of reasoning by analogy [10, 12, 22, 24]. Similarity on t-norm based fuzzy logic was introduced by Castro [5] with the intent of measuring the similarity degree of each couple of truth of propositions in fuzzy systems, which is a generalization of equivalence on classical logic. Considering that random experiments may also follow the rules of other fuzzy systems, the notion of similarity has been extended to various logic systems such as predicate **BL** [19], Łukasiewicz [30] and their non-commutative cases [13]. Although these way can be expanded the scope of similarity, they both have as codomain the closed unit interval [0,1]. However, fuzzy logical algebras with similarities are not Universal Algebra and hence they do not automatically induce an assertional logic. To present a unified approach to similarity and introduce in the many valued context a deduction apparatus able to reason by analogy in a logical and algebraic setting, a new approach to similarities on MV-algebras was introduced by Gerla and Leuştean [16], where they added a binary operation S to the language of MV-algebras as a similarity satisfying some basic properties of similarity. The resulting algebras structures were so-called similarity MV-algebras. This approach generalizes the similarity, as a function on the algebra taking values in the interval [0,1] with the addition property, as well as Hájek's approach to fuzzy logic with very true in [20]. Moreover, Gerla and Leuştean presented an algebraizable logic, and its equivalent algebraic semantics is precisely the variety of similarity MV-algebras, and proved the completeness of them. Recently, Wang introduced in [35] similarity MTL-algebras, which provide a more general algebraic foundation for the similarity degree of each couple of truth degrees of propositions in **MTL**, and gave some characterizations of representable similarity MTL-algebras.

In this paper, we will extend similarity to monadic MTL-algebras for providing a more general algebraic foundation for the similarity degree of each couple of truth degrees of predicate variables in monadic monoidal t-norm based predicate logic. The main focus of existing research about similarity is on MV-algebras [16], DH-algebras [1], Łukasiewicz-Moisil algebras [7] and MTL-algebras [35]. All the above mentioned algebraic structures are the algebraic semantics of t-norm based on fuzzy propositional logic. However, there is no research about algebras of similarity in monadic fuzzy predicate t-norm based fuzzy logic so far. Therefore, it is interesting to study similarity on monadic MTL-algebras for treating a variant of the concept of similarities within the framework of Universal Algebra and provide a sold algebraic foundation for the similar degree of each couple of truth degrees of predicate variables in monadic t-norm fuzzy based predicate logic. These are the motivations for us to investigate similarity on monadic MTL-algebras.

The paper is organized as follows. In Section 2, we review some basic definitions and results about monadic MTL-algebras. In Section 3, we introduce similarity monadic MTL-algebras and study some of their related algebraic properties. In Section 4, we introduce and investigate similarity filters in the similarity monadic MTL-algebras and give some characterizations of representable similarity monadic MTL-algebras. In Section 5, we introduce the the logic of similarity monadic MTL-algebras and prove the soundness and completeness of them.

2 Preliminaries

In this section, we review some basic results on MTL-algebras and their related monadic algebraic structures.

Definition 2.1. [11] *An MTL-algebra is an algebraic structure* $(L, \cup, \cap, \otimes, \to, 0, 1)$ *with four binary operations and two constants $0, 1$ such that:*
  (1) $(L, \cup, \cap, 0, 1)$ *is a bounded lattice,*
  (2) $(L, \otimes)$ *is a commutative monoid,*
  (3) $p \otimes q \leq r$ *if and only if* $p \leq q \to r$,
  (4) $(p \to q) \cup (q \to r) = 1$,
for any $p, q, r \in L$.

In what follows, by L we denote the universe of an MTL-algebra $(L, \cup, \cap, \otimes, \to, 0, 1)$. In any MTL-algebra L, we define

$$\neg p = p \to 0, \neg\neg p = \neg(\neg p), p^0 = 1 \text{ and } p^n = p^{n-1} \odot p \text{ for } n \geq 1.$$

Proposition 2.2. [35] *Let $(L, \cup, \cap, \otimes, \to, 0, 1)$ be an MTL-algebra. Then the following properties are valid, for all $p, q, r \in L$:*
  (1) $p \cup q \leq (p \to q) \to q$ *(in particular $p \leq \neg\neg p$),*
  (2) $p \to q \leq p \otimes r \to q \otimes r$,
  (3) $(p \to q) \otimes (q \to r) \leq (p \to r)$,
  (4) *If $p \leq q$, then $p \otimes r \leq q \otimes r, r \to p \leq r \to q$ and $q \to r \leq p \to r$,*
  (5) $p \to (q \to r) = q \to (p \to r) = (p \otimes q) \to r$,
  (6) $p \leftrightarrow q \leq (q \leftrightarrow r) \leftrightarrow (p \leftrightarrow r)$,
  (7) *If $\cap_{i \in I} p_i, \cup_{i \in I} q_i$ and $\cap_{i \in I}(q_i \to p_i)$ exist, then $\cap_{i \in I}(p_i \to q_i) \leq \cup_{i \in I} p_i \to \cup_{i \in I} q_i$,*
  (8) *If $\cap_{i \in I} p_i, \cap_{i \in I} q_i$ and $\cap_{i \in I}(q_i \to p_i)$ exist, then $\cap_{i \in I}(p_i \to q_i) \leq \cap_{i \in I} p_i \to \cup_{i \in I} q_i$.*

In the sequel, we recall some representations of MTL-algebras. In order to do so, we start from recalling filters of MTL-algebras in [3].

A nonempty subset F of an MTL-algebra L is called a *filter* if it satisfies: (1) $1 \in F$; (2) $p \in F$ and $p \to q \in F$ imply $q \in F$. A filter F of L is called a *proper filter* if $F \neq L$. Unless otherwise explicitly stated, filters are assumed to be proper. A proper filter F of L is called a *maximal filter* if it is not contained in any proper filter of L. A proper filter F of L is called a *prime filter* if for each $p, q \in L$ and $p \vee q \in F$, imply $p \in F$ or $q \in F$. A prime filter F is said to be *minimal* if it is a minimal element in the set of prime filters of L ordered by inclusion. Moreover, we denote by $\langle X \rangle$ be the filter generated by a nonempty subset X of L. Clearly

$$\langle X \rangle = \{p \in L \mid p \geq p_1 \otimes p_2 \otimes \cdots \otimes p_n, \text{ for some } n \in N \text{ and some } p_i \in X\}.$$

In particular, the principal filter generated by an element $p \in L$ is

$$\langle p \rangle = \{q \in L \mid q \geq p^n\}.$$

If F is a filter and $p \in L$, then

$$\langle F \cup \{p\} \rangle = \{q \in L \mid q \geq f \otimes p^n, \text{ for some } f \in F\}.$$

Definition 2.3. [26] *An algebraic structure $(L, \cap, \cup, \otimes, \to, 0, 1, \forall, \exists)$ is said to be a monadic residuated lattice if $(L, \cap, \cup, \otimes, \to, 0, 1)$ is a residuated lattice and in addition $\forall$ and $\exists$ satisfy the following identities:*
  ($\forall 1$) $\forall p \to p = 1$,
  ($\exists 1$) $p \to \exists p = 1$,
  ($\forall 2$) $\forall (p \to \exists q) = \exists p \to \exists q$,
  ($\forall 3$) $\forall (\exists p \to q) = \exists p \to \forall q$,
  ($\forall 4$) $\forall (p \cup \exists q) = \forall p \cup \exists q$,
  ($\forall 5$) $\forall \forall p = p$,
  ($\exists 2$) $\exists \forall p = \forall p$,
  ($\exists 3$) $\exists (p \otimes p) = \exists p \otimes \exists p$,
  ($\exists 4$) $\exists (\exists p \otimes \exists q) = \exists p \otimes \exists q$,
for any $p, q \in L$.

Remark 2.4. [36] *($\exists 1$), ($\forall 5$) and ($\exists 4$) are redundant in monadic residuated lattices.*

MTL-algebras form a subclass of residuated lattices, so we apply the axioms of Definition 2.3 from residuated lattices to MTL-algebras and obtain monadic MTL-algebras.

Definition 2.5. [36] *A monadic MTL-algebra is a structure structure* $(L, \cup, \cap, \otimes, \to, 0, 1, \forall, \exists)$ *in which* $(L, \cup, \cap, \otimes, \to, 0, 1)$ *is an MTL-algebra,* $\forall$ *and* $\exists$ *are two unary operations on L, and satisfying the conditions:* $(\forall 1), (\forall 2), (\forall 3), (\forall 4), (\exists 2), (\exists 3)$.

In the sequel, by $(L, \forall, \exists)$ we denote the universe of a monadic MTL-algebra $(L, \cap, \cup, \otimes, \to, 0, 1, \forall, \exists)$.

Proposition 2.6. [26] *Let $(L, \forall, \exists)$ be a monadic MTL-algebra. Then the following properties hold: for any $p, q \in L$,*
  (1) $\forall (p \to q) \to (\forall p \to \forall q) = 1$,
  (2) $\forall (p \cap q) = \forall p \cap \forall q$,
  (3) $p \to \exists p = 1$,
  (4) $\exists (\exists p \to q) \to (\exists p \to \exists q) = 1$,
  (5) $\forall (p \to \forall q) = \exists p \to \forall q$,
  (6) $\forall (\forall p \to q) = \forall p \to \forall q$,
  (7) $\exists (p \cap \exists q) = \exists p \cap \exists q$,
  (8) $\forall \exists p = \exists p$,
  (9) $\exists \forall p = \forall p$,
  (10) $\exists (p \cup q) = \exists p \cup \exists q$,
  (11) $\forall (\forall p \otimes \forall q) = \forall p \otimes \forall q$,
  (12) $\forall (\forall p \to \forall q) = \forall p \to \forall q$,
  (13) $\forall L = \exists L = L_{\forall \exists}$, where $L_{\forall \exists} = \{p \in L | \forall p = p\} = \{p \in L | \exists p = p\}$,
  (14) $L_{\forall \exists}$ is a subalgebra of L.

Definition 2.7. [4] *A monadic BL-algebra is an algebraic structure* $(L, \cup, \cap, \otimes, \to, 0, 1, \forall, \exists)$ *in which* $(L, \cup, \cap, \otimes, \to, 0, 1)$ *is a BL-algebra,* $\forall$ *and* $\exists$ *are two unary operations on L, and satisfying the conditions:* $(\forall 1), (\exists 3)$ *and*
  ($\forall 6$) $\forall (p \to \forall q) = \exists p \to \forall q$,
  ($\forall 7$) $\forall (\forall p \to q) = \forall p \to \forall q$,
  ($\forall 8$) $\forall (\exists p \cup q) = \exists p \cup \forall q$,
for any $p, q \in L$.

Theorem 2.8. [36] *Let L be an MTL-algebra and $\forall : L \to L$ and $\exists : L \to L$ be two unary operations on L. Then the sets*

$$G = \{(\forall 1), (\forall 2), (\forall 3), (\forall 4), (\exists 2), (\exists 3)\},$$
$$W = \{(\forall 1), (\forall 6), (\forall 7), (\forall 8), (\exists 3)\}$$

are equivalent for an MTL-algebra L.

3 Similarity monadic MTL-algebras

In this section, we introduce the notion of similarity monadic MTL-algebras and study some of their related algebraic properties.

Definition 3.1. *A* similarity monadic MTL-algebra *is a quadruple* $(L, \forall, \exists, S)$, *where* $(L, \forall, \exists)$ *is a monadic MTL-algebra and* $S : L \times L \longrightarrow L$ *is a binary operation on* L *such that the following properties hold for all* $p, q, r \in L$,

(S1) $S(p,p) = 1$,
(S2) $S(p,q) = S(q,p)$,
(S3) $S(p,q) \otimes S(q,r) \leq S(p,r)$,
(S4) $p \otimes S(p,q) \leq q$,
(S5) $S(p \leftrightarrow q, 1) \leq S(p,r) \leftrightarrow S(q,r)$,
(S6) $\forall S(p,q) \leq S(\forall p, \forall q)$,
(S7) $\exists S(\exists p, \exists q) \leq S(\exists p, \exists q)$.

It is noted that the binary operation $S : L \times L \longrightarrow L$ *which satisfies (S1)-(S5) will be called a similarity on an MTL-algebra* L *[35]. If* S *and* T *are two similarities on a monadic MTL-algebra* $(L, \forall, \exists)$, *then we define*

$$S \leq T \text{ if and only if } S(p,q) \leq T(p,q), \text{ for any } p, q \in L.$$

Remark 3.2. *The class of monadic MTL-algebras is a variety of algebras. We can present the whole axioms of similarity monadic MTL-algebras with equality, so the class of all similarity monadic MTL-algebras can from a variety.*

Proposition 3.3. *Let* $(L, \forall, \exists, S)$ *be a similarity monadic MTL-algebra. Then the following statements are equivalent:*

(1) $S(p,q) = \forall S(p,q)$,
(2) $S(p,q) = \exists S(p,q)$.

Proof. (1) $\Rightarrow$ (2) If $S(p,q) = \forall S(p,q)$, then by Proposition 2.6(9), we have

$$S(p,q) = \forall S(p,q) = \exists \forall S(p,q) = \exists S(p,q),$$

which implies $S(p,q) = \exists S(p,q)$.

(2) $\Rightarrow$ (1) If $S(p,q) = \exists S(p,q)$, then by Proposition 2.6(8), we have

$$S(p,q) = \exists S(p,q) = \forall \exists S(p,q) = \forall S(p,q),$$

which implies $S(p,q) = \forall S(p,q)$. □

Example 3.4. (1) *Let* $(L, \forall, \exists)$ *be a monadic MTL-algebra and* $S : L \times L \longrightarrow L$ *be defined by*

$$\triangle(p,q) := \begin{cases} 1 & p = q \\ 0 & p \neq q \end{cases}$$

for all $p, q \in L$. Then $(L, \forall, \exists, \triangle)$ is a similarity monadic MTL-algebra.

(2) Let $(L, \forall, \exists)$ be a monadic MTL-algebra and

$$I(p,q) := p \leftrightarrow q.$$

Then I is a similarity on $(L, \forall, \exists)$. The axiom (S5) follows by Proposition 2.2. Also, by Propositions 2.6(1) and (2), we have

$$\begin{aligned} \forall I(p,q) &= \forall (p \leftrightarrow q) \\ &= \forall ((p \to q) \cap (q \to p)) \\ &= \forall (p \to q) \cap \forall (q \to p) \\ &\leq (\forall p \to \forall q) \cap (\forall q \to \forall p) \\ &= \forall p \leftrightarrow \forall q \\ &= I(\forall p, \forall q). \end{aligned}$$

Then (S6) holds.

For axiom (S7) by Proposition 2.2 and Propositions 2.6(4) and (7), we have:

$$\begin{aligned} \exists I(\exists p, \exists q) &= \exists (\exists p \leftrightarrow \exists q) \\ &= \exists ((\exists p \to \exists q) \cap (\exists q \to \exists q)) \\ &\leq \exists (\exists p \to \exists q) \cap \exists (\exists q \to \exists q) \\ &= (\exists p \to \exists q) \cap (\exists q \to \exists p) \\ &= \exists p \leftrightarrow \exists q \\ &= I(\exists p, \exists q). \end{aligned}$$

Thus $(L, \forall, \exists, I)$ is a similarity monadic MTL-algebra.

Remark 3.5. (1) It is easy to checked that if S is a similarity on a monadic MTL-algebra $(L, \forall, \exists)$, then $\triangle \leq S \leq I$, which implies that I and $\triangle$ are extremal similarities on a monadic MTL-algebra $(L, \forall, \exists)$, respectively.

(2) Zahiri and Borumand Saeid introduced in [44] a similarity monadic BL-algebra as a quadruple $(L, \forall, \exists, S)$ that satisfies the axioms: (S1) − (S6) and

$$(S7)' \quad \exists S(p, q) \leq S(\exists p, \exists q),$$

and show that $(L, \forall, \exists, I)$ in Example 3.4(2) is a similarity monadic BL-algebra. Indeed, they have

$$\begin{aligned}
\exists I(p,q) &= \exists(p \leftrightarrow q) \\
&= \exists((p \to q) \cap (q \to p)) \\
&\leq \exists(p \to q) \cap \exists(q \to p) \\
&\leq \exists(\exists p \to q) \cap \exists(\exists q \to p) \\
&= (\exists p \to \exists q) \cap (\exists q \to \exists p) \\
&= \exists p \leftrightarrow \exists q \\
&= I(\exists p, \exists q).
\end{aligned}$$

However, in the proof of $(S7)'$, the inequality

$$\exists(p \to q) \cap \exists(q \to p) \leq \exists(\exists p \to q) \cap \exists(\exists q \to p),$$

is not true in general, since

$$\exists(p \to q) \leq \exists(\exists p \to q)$$

is not hold in any monadic BL-algebra in general. Otherwise, $\forall = \exists = id_L$, where id_L is a unary identity operator on a BL-algebra L.

Now we give an example of a finite similarity monadic MTL-algebra.

Example 3.6. let $L = \{0, h, m, n, 1\}$, with $0 < h < m < 1, 0 < h < n < 1$. We define $\otimes$ and $\to$ are as follows:

$\otimes$	0	h	m	n	1
0	0	0	0	0	0
h	0	h	h	h	h
m	0	h	m	h	m
n	0	h	h	n	n
1	0	h	m	n	1

$\to$	0	h	m	n	1
0	1	1	1	1	1
h	0	1	1	1	1
m	0	n	1	n	1
n	0	m	m	1	1
1	0	h	m	n	1

Then $(L, \cap, \cup, \otimes, \to, 0, 1)$ is a BL-algebra and hence an MTL-algebra. Defining $\forall$ and $\exists$ are as follows,

p	0	h	m	n	1
$\forall p$	0	h	h	h	1

p	0	h	m	n	1
$\exists p$	0	h	1	1	1

It is verified that $(L, \forall, \exists)$ is a monadic MTL-algebra and . Now, we define S as follow:

$$S(0,0) = S(h,h) = S(m,m) = S(n,n) = S(1,1) = 1,$$
$$S(0,h) = S(h,0) = S(0,m) = S(m,0) = S(0,n) = S(n,0) = S(0,1) = S(1,0) = 0,$$
$$S(h,1) = S(1,h) = h, S(m,1) = S(1,m) = m, S(n,1) = S(1,n) = n,$$
$$S(h,m) = S(m,h) = n, S(h,n) = S(n,h) = m, S(m,n) = S(n,m) = h.$$

Then $(L, \forall, \exists, S)$ is a similarity monadic MTL-algebra. However, $(L, \forall, \exists, S)$ is not a similarity monadic BL-algebra in [44]. Indeed, $(S7)'$ is not hold in the case,

$$S(\exists h, \exists m) = S(h,1) = h, \quad \exists S(h,m) = \exists n = 1,$$

and hence

$$\exists S(h,m) \not\leq S(\exists h, \exists m).$$

Proposition 3.7. Let $(L, \forall, \exists, S)$ be a similarity monadic MTL-algebra. Then the following hold, where $S(p,q) = I(p,q)$ and for any $p,q,r,u \in L$,
  (1) $S(p,q) = 1$ if and only if $p = q$,
  (2) $S(p,r) \otimes S(q,r) \leq S(p,q)$,
  (3) if $p,q \in [r,u]$, then $S(r,u) \leq S(p,q)$,
  (4) $S(p,q) \leq S(p \otimes r, q \otimes r)$,
  (5) $S(p,r) \otimes S(q,u) \leq S(p \otimes q, r \otimes u)$,
  (6) $S(p,r) \otimes S(q,u) \leq S(q \to p, u \to r)$,
  (7) $S(p,r) \cap S(q,u) \leq S(p \cap q, r \cap u)$,
  (8) $S(p,r) \cap S(q,u) \leq S(p \cup q, r \cup u)$,
  (9) $S(1,p) = p$ and $S(0,p) = \neg p$.

Proof. The proof of parts $(1), (8), (9)$ are clear, so we omit them.
  (2) By Propositions 2.2(3) and (4), we have

$$\begin{aligned} S(p,q) &= (p \to q) \cap (q \to p) \\ &\geq [(r \to p) \otimes (q \to r)] \cap [(r \to q) \otimes (p \to r)] \\ &\geq [(r \to p) \cap ((r \to q)] \otimes [(r \to p) \cap (p \to r)] \otimes [(q \to r) \cap (r \to q)] \otimes \\ &\quad [(q \to r) \cap (p \to r)] \\ &= [(r \to p) \cap (p \to r)] \otimes [(q \to r) \cap (r \to q)] \\ &= S(p,r) \otimes S(r,q). \end{aligned}$$

  (3) If $p,q \in [r,u]$, then $p \to q \geq p \to r \geq u \to r$, and hence

$$q \to p \geq q \to r \geq u \to r,$$

which implies $S(p,q) \geq u \to r \geq S(r,u)$.
  (4) By Proposition 2.2(2), we have $p \to q \leq (p \otimes r) \to (q \otimes r)$, and hence

$$q \to p \leq (q \otimes r) \to (p \otimes r),$$

which implies $S(p,q) \leq S(p \otimes r, q \otimes r)$.

(5) By (2),(4), we have $S(p \otimes q, r \otimes u) \geq S(p \otimes q, r \otimes q) \otimes S(r \otimes q, r \otimes u) \geq S(p,r) \otimes S(q,u)$.

(6) We show that $(u \to p) \otimes (q \to r) \leq (r \to u) \to (q \to p)$. This is equivalent to $(u \to p) \otimes (r \to u) \otimes (q \to r) \leq (q \to p)$, so by Proposition 2.2(3), we have $(u \to p) \otimes (r \to u) \otimes (q \to r) \otimes q \leq p$. Hence $(p \to u) \otimes (r \to q) \leq (q \to p) \to (r \to u)$. Thus

$$\begin{aligned} S(q \to p, r \to u) &\geq [(u \to p) \otimes (q \to r)] \cap [(p \to u) \otimes (r \to q)] \\ &\geq [(u \to p) \otimes (q \to r)] \cap [(p \to u) \otimes (r \to q)] \cap [(u \to p) \otimes (r \to q)] \\ &\quad \cap [(p \to u) \otimes (q \to r)] \\ &\geq [(u \to p) \cap (p \to u)] \otimes [(q \to r) \cap (r \to q)] \\ &= S(p,u) \otimes S(q,r). \end{aligned}$$

(7) By Proposition 2.2(7), we have $(u \to p) \cap (r \to q) \leq (u \cap r) \to (p \cap q)$, and $(p \to u) \cap (q \to r) \leq (p \cap q) \to (u \cap r)$. So $S(p,u) \cap S(q,r) \leq S(p \cap q, u \cap r)$. □

Definition 3.8. *Let X be a set. An L-fuzzy subset of X is a function $f: X \longrightarrow L$ and an L-similarity on X is a fuzzy subset $E: X \times X \longrightarrow L$ of $X \times X$ such that the following properties hold for any $p,q,r \in L$,*

(E1) $E(p,p) = 1$,
(E2) $E(p,q) = E(q,p)$,
(E3) $E(p,q) \otimes E(q,r) \leq E(p,r)$.

An L-fuzzy subset $f: X \longrightarrow L$ is extensional with respect to E if for all $p, q \in X$,

(E4) $f(p) \otimes E(p,q) \leq f(q)$.

Example 3.9. *In Example 3.6, if $X = \{h, m, n, 1\}$ and $S(p,q) = I(p,q)$, then $f(p) = p^3$ is an L-fuzzy subset of X. Also for all $p, q \in X$, we have $f(p) \otimes I(p,q) \leq f(q)$, so f is an extensional with respect to I.*

Remark 3.10. *Let $(L, \forall, \exists, S)$ be a similarity monadic MTL-algebra, X be a set and $f: X \longrightarrow L$ be an L-subset of X. If we can define*

$$E: X \times X \longrightarrow L,$$
$$E(p,q) := S(f(p), f(q)),$$

then it is clear that E is an L-similarity on X and f is extensional with respect to E.

Let $(L, \forall, \exists)$ be a monadic MTL-algebra and X a set. Then $(L^X, \forall^X, \exists^X)$ is also a monadic MTL-algebra with point wise operations, where

$$0(p) = 0, \quad 1(p) = 1,$$
$$(f \cup g)(p) = f(p) \cap g(p), \quad (f \cup g)(p) = f(p) \cup g(p),$$
$$(f \otimes g)(p) = f(p) \otimes g(p), \quad (f \to g)(p) = f(p) \to g(p),$$
$$\forall^X(f) = \forall \circ f, \quad \exists^X(f) = \exists \circ f.$$

Proposition 3.11. *Let $\mathcal{S}_X = \{S_X : p \in X\}$ be a family of similarities on $(L, \forall, \exists)$. Then there is an one-to-one correspondence between $\mathcal{S}_X$ and a family of similarities on $(L^X, \forall^X, \exists^X)$.*

Proof. Let $\mathcal{S}_X = \{S_X : p \in X\}$ be a family of similarities on $(L, \forall, \exists)$. Then we define a similarity on a monadic MTL-algebra $(L^X, \forall^X, \exists^X)$ by

$$S_{\mathcal{S}_X}(f, g)(p) := S_X(f(p), g(p)),$$

for all $f, g \in L^X$ and $p \in X$. Clearly, $(L^X, \forall^X, \exists^X, S_{\mathcal{S}_X})$ is a similarity monadic MTL-algebra.

Conversely, if

$$K : (L^X, \forall^X, \exists^X) \times (L^X, \forall^X, \exists^X) \longrightarrow (L^X, \forall^X, \exists^X)$$

is a similarity on the monadic MTL-algebra $(L^X, \forall^X, \exists^X)$, then for $a \in L$, we denote by f_a the subset $f_a(p) = a$ for all $p \in X$. So we can define a similarity on $(L, \forall, \exists)$ by

$$K_p : (L, \forall, \exists) \times (L, \forall, \exists) \longrightarrow (L, \forall, \exists),$$

$$K_p(a, b) := K(f_a, f_b)(p),$$

for $p \in X$. Thus we get a family of similarities $\{K_p \mid p \in X\}$ on a monadic MTL-algebra $(L, \forall, \exists)$. $\square$

Proposition 3.12. *Let $(L, \forall, \exists, S)$ be a similarity monadic MTL-algebra. Then the following statements are equivalent:*
  (1) $S(p, 1) = p$,
  (2) $S(p, q) = p \leftrightarrow q$,
for all $p, q \in L$.

Proof. (1) $\Rightarrow$ (2) If $S(p, q) = p \leftrightarrow q$, then $S(p, 1) = p$.
  (2) $\Rightarrow$ (1) By (S1) and (S5), we have

$$p \leftrightarrow q = S(p \leftrightarrow q, 1) \leq S(p, q) \leftrightarrow S(q, q) = S(p, q) \leftrightarrow 1 = S(p, q).$$

On the other hand by (S2) and (S4), we have $S(p, q) \leq p \leftrightarrow q$. $\square$

4 Representations of similarity monadic MTL-algebras

In this section, we introduce similarity monadic filter of similarity monadic MTL-algebras and give some characterizations of representable similarity monadic MTL-algebras by them.

Definition 4.1. *Let $(L, \forall, \exists, S)$ be a similarity monadic MTL-algebra. Then F is called a* similarity monadic filter *of $(L, \forall, \exists, S)$ if F is a monadic filter of $(L, \forall, \exists)$ and*

$$S(\forall p, \forall q) \in F \text{ whenever } p, q \in F.$$

A similarity monadic filter is called prime *if $p \to q \in F$ or $q \to p \in F$, for all $p, q \in L$.*
It is noted that the filter F of L is called a similarity filter of a similarity MTL-algebra if F such that $x, y \in F$, then $S(x, y) \in F$, for all $x, y \in F$.

Example 4.2. *In Example 3.6, $F = \{1\}$ is a similarity monadic filter of $(L, \forall, \exists, I)$. However, F is not a similarity monadic prime filter. If we taking $G = \{h, m, n, 1\}$, then G is a similarity monadic prime filter of $(L, \forall, \exists, S)$.*

Proposition 4.3. *Let $(L, \forall, \exists, S)$ be a similarity monadic MTL-algebra and F be a similarity monadic filter of $(L, \forall, \exists, S)$. Then the following statements are equivalent:*
  (i) *F is a similarity monadic filter of $(L, \forall, \exists, S)$,*
  (ii) *$S(\forall p, 1) \in F$, for all $p \in F$.*

Proof. By (S3), we have $S(\forall p, 1) \otimes S(\forall q, 1) \leq S(\forall p, \forall q)$. So F is a similarity monadic filter if and only if $S(\forall p, 1) \in F$, whenever $p \in F$. □

Proposition 4.4. *Let $(L, \forall, \exists, S)$ be a similarity monadic MTL-algebra and F be a similarity monadic filter of $(L, \forall, \exists, S)$. Then $\equiv_F$ is a congruence on a similarity monadic MTL-algebra $(L, \forall, \exists, S)$.*

Proof. Let $p_1, p_2, q_1, q_2 \in L$ be such that $p_1 \equiv_F p_2$ and $q_1 \equiv_F q_2$. Then $p_1 \leftrightarrow p_2 \in F$, $q_1 \leftrightarrow q_2 \in F$. Since F is a similarity monadic filter, we have $S(\forall(p_1 \leftrightarrow p_2), 1) \in F$ and $S(\forall(q_1 \leftrightarrow q_2), 1) \in F$. Using (S2), (S3) and the definition of monadic filter, we have

$$S(\forall(p_1 \leftrightarrow p_2), 1) \otimes S(\forall(q_1 \leftrightarrow q_2), 1) \leq S(\forall(p_1 \leftrightarrow p_2), \forall(q_1 \leftrightarrow q_2)) \in F.$$

So by (S5), we have $\forall(p_1 \leftrightarrow q_1) \leftrightarrow \forall(p_2 \leftrightarrow q_2) \in F$. Thus $(p_1 \leftrightarrow q_1) \equiv_F (p_2 \leftrightarrow q_2)$. □

Proposition 4.5. Let $(L, \forall, \exists, S)$ be a similarity monadic MTL-algebra and $F \subseteq L$ be a similarity monadic filter of $(L, \forall, \exists, S)$. Then $(L/\equiv_F, \forall_F, \exists_F)$ is also a similarity monadic MTL-algebra.

Proof. We define
$$S_F : L/\equiv_F \times L/\equiv_F \longrightarrow L/\equiv_F,$$
$$S_F([p]_F, [q]_F) := [S(p, q)]_F,$$
$$\forall_F : L/\equiv_F \longrightarrow L/\equiv_F,$$
$$\forall_F([p]_F) := [\forall p]_F,$$
$$\exists_F : L/\equiv_F \longrightarrow L/\equiv_F,$$
$$\exists_F([p]_F) := [\exists p]_F$$

for all $p, q \in L$.

(S1) $S_F([p]_F, [q]_F) = [S(p, p)]_F = [1]_F$,
(S2) $S_F([p]_F, [q]_F) = [S(p, q)]_F = [S(q, p)]_F = S_F([q]_F, [p]_F)$,
(S3) $S_F([p]_F, [q]_F) \otimes S_F([q]_F, [r]_F) = [S(p, q) \otimes S(q, r)]_F \leq [S(p, r)]_F = S_F([p]_F, [r]_F)$,
(S4) $[p]_F \otimes S_F([p]_F, [q]_F) = [p]_F \otimes [S(p, q)]_F \leq [q]_F$,
(S5) $S_F([p]_F \leftrightarrow [q]_F, 1) = [S(p \leftrightarrow q, 1)]_F \leq [S(p, r)]_F \leftrightarrow [S(q, r)]_F = S_F([p]_F, [r]_F) \leftrightarrow S_F([q]_F, [r]_F)$,
(S6) $\forall_F S_F([p]_F, [q]_F) = \forall_F [S(p, q)]_F \leq [S(\forall p, \forall q)]_F = S_F(\forall_F [p]_F, \forall_F [q]_F)$,
(S7) $\exists_F S_F([p]_F, [q]_F) = \exists [S(p, q)]_F \leq [S(\exists p, \exists q)]_F = S_F(\exists_F [p]_F, \exists_F [q]_F)$. □

Definition 4.6. A similarity monadic MTL-algebra is called **representable** *if it is a subdirect product of linearly ordered similarity monadic MTL-algebras.*

Theorem 4.7. *Let $(L, \forall, \exists, S)$ be a similarity monadic MTL-algebra. Then the following statements are equivalent:*
  (i) $(L, \forall, \exists, S)$ *is representable,*
  (ii) $S(\forall(p \to q), 1) \cup \forall(q \to p) = 1$, *for all $p, q \in L$,*
  (iii) $p \cup q = 1$ *implies $\forall p \cup S(\forall q, 1) = 1$, for all $p, q \in L$,*
  (iv) *any minimal monadic prime filter is a similarity monadic filter of $(L, \forall, \exists, S)$.*

Proof. $(i) \Rightarrow (ii)$ In any linearly ordered similarity monadic MTL-algebra, we have
$$S(\forall p \to \forall q) \cup (\forall q \to \forall p) = 1.$$

$(ii) \Rightarrow (iii)$ If $p \cup q = 1$, then by Proposition 2.2 (1), we have $p \to q = q$, $q \to p = p$. Using (ii), $\forall p \cup S(\forall q, 1) = 1$.

$(iii) \Rightarrow (iv)$ If $F \subseteq L$ is a minimal monadic prime filter of $(L, \forall, \exists)$ and $p \in F$, then there exists $r \in L$ such that $\forall p \cap \forall r = 1$ and $r \notin F$. By (iii), we have $S(\forall p, 1) \cup \forall r = 1 \in F$. Since $r \notin F$ and F is monadic prime filter, $S(\forall p, 1) \in F$.

$(iv) \Rightarrow (i)$ Let $(L, \forall, \exists, S)$ be a similarity monadic MTL-algebra and Λ be a set of all the minimal monadic prime filter of $(L, \forall, \exists)$. Clearly, L is a subdirect product of family of the $\{L/\equiv_F : F \in \Lambda\}$. Let $\varphi : L \longrightarrow \prod_{F \in I} L/\equiv_F$ be the subdirect product. By (iv), any $F \in I$ is a similarity monadic filter of $(L, \forall, \exists, S)$, hence by Proposition 4.5, we have $(L/\equiv_F, \forall_F, \exists_F, S_F)$ is a similarity monadic MTL-algebra. It is clear that φ is a representation of $(L, \forall, \exists, S)$ as a subdirect product of the family $(L/\equiv_F, \forall_F, \exists_F, S_F)$. □

5 The logic of similarity monadic MTL-algebras

Hájek proved in [20] that monadic predicate basic logic $\mathbf{mBL}_\forall$ is equivalent to S5-like modal fuzzy logic $\mathbf{S5(BL)}$, which is a logic $\mathbf{BL}$ together with the following axioms (ν is a propositional combination of formulas beginning by $\Box$ and $\Diamond$)

($\Box$1) $\Box\varphi \Rightarrow \varphi$,

($\Diamond$1) $\varphi \Rightarrow \Diamond\varphi$,

($\Box$2) $\Box(\nu \Rightarrow \varphi) \Rightarrow (\nu \Rightarrow \Box\varphi)$,

($\Diamond$2) $\Box(\varphi \Rightarrow \nu) \Rightarrow (\Diamond\varphi \Rightarrow \nu)$,

($\Box$3) $\Box(\nu \sqcup \varphi) \Rightarrow (\nu \sqcup \Box\varphi)$,

($\Diamond$3) $\Diamond(\varphi \& \varphi) \equiv \Diamond\varphi \& \Diamond\varphi$,

closed under Modus Ponens $\mathbf{MP}$: $\varphi, \varphi \Rightarrow \psi \vdash \psi$ and Necessitation Rule $\mathbf{Nec}$: $\varphi/\Box\varphi$. Subsequently, Castaño et. al introduced monadic BL-algebras and proved that are the equivalent algebraic semantics of the logic $\mathbf{mBL}_\forall$ (and $\mathbf{S5(BL)}$) [4].

As a consequence of the algebraization of $\mathbf{S5(BL)}$ by monadic BL-algebras, Castaño et. al also gave a simplified set of axioms for this calculus, which is the propositional case for the axiomatization of $\mathbf{S5(BL)}$. Here they defined a calculus $\mathbf{S5'(BL)}$ whose axiom schemata are the ones for $\mathbf{BL}$ together with the following axiom schemata:

(M1) $\Box\varphi \Rightarrow \varphi$,

(M2) $\varphi \Rightarrow \Diamond\varphi$,

(M3) $\Box(\Box\varphi \Rightarrow \psi) \Rightarrow (\Box\varphi \Rightarrow \Box\psi)$,

(M4) $\Box(\varphi \Rightarrow \Box\psi) \Rightarrow (\Diamond\varphi \Rightarrow \Box\psi)$,

(M5) $\Box(\Box\varphi \sqcup \psi) \Rightarrow (\Box\varphi \sqcup \Box\psi)$,

(M6) $\Diamond(\varphi \& \varphi) \equiv \Diamond\varphi \& \Diamond\varphi$

and closed under $\mathbf{MP}$: $\varphi, \varphi \Rightarrow \psi \vdash \psi$ and $\mathbf{Nec}$: $\varphi/\Box\varphi$, and showed in [4] that $\mathbf{S5'(BL)}$ is sound and complete with respect to the variety of monadic BL-algebras.

It is worth noticing that monadic BL-algebra and monadic MTL-algebra have the same axioms by Theorem 2.8. So, we can analogously define the modal fuzzy logic **S5(MTL)** as the logic **MTL** together with the axioms ($\Box$1), ($\Box$2), ($\Box$3), ($\Diamond$1),($\Diamond$2), ($\Diamond$3), and their corresponding modal fuzzy propositional logic **S5'(MTL)**, which is the logic **MTL** together with the axioms (M1), (M2), (M3), (M4), (M5) and (M6). Along the same line as that in [20] (see [12], Theorem 6) and [19] (see [11], Remark 8.3.16), we can also prove that the modal fuzzy logic **S5(MTL)** is equivalent to monadic predicate monoidal t-norm based logic **mMTL**$_\forall$.

In this section, adapting for the propositional case the axiomatization of similarity monadic MTL-algebras, we introduce the logic **SMMTL** and show that is sound and complete with respect to the variety of similarity monadic MTL-algebras.

The language of **SMMTL** consists of countably many proposition variables ($v_1, v_2, ...$), the constant $\bar{0}$, the unary operators $\Box$, $\Diamond$, the binary operators $\sqcup, \sqcap, \&, \Rightarrow$, a binary logic connective $\Leftrightarrow$, the auxiliary symbol '(/and')/.

Formulas are defined inductively : $\bar{0}$ is a formula; if φ and ψ are formulas, then so are ($\varphi \sqcap \psi$), ($\varphi \sqcup \psi$), ($\varphi \& \psi$), ($\varphi \Rightarrow \psi$), ($\varphi \Leftrightarrow \psi$), ($\Box \varphi$) and ($\Diamond \varphi$). And we state that:

$$\varphi \leftrightarrow \psi := (\varphi \Rightarrow \psi) \sqcap (\psi \Rightarrow \varphi).$$

In order to avoid unnecessary brackets, we agree on the following priority rules:
- unary operators always take precedence over binary ones, while,
- among the binary operators, & has the highest priority; furthermore $\sqcup$ and $\sqcap$ take precedence over $\Rightarrow$,
- the outermost brackets are not write.

The axiom of **SMMTL** are defined as follows:
(I). Any axioms of **S5'MTL** is an axiom of **SMMTL**
(II). A formula which has one of the following forms is an axiom (where φ and ψ are arbitrary formulas):
(s1) $\varphi \Leftrightarrow \varphi$,
(s2) $(\varphi \Leftrightarrow \psi) \Rightarrow (\psi \Leftrightarrow \varphi)$,
(s3) $(\varphi \Leftrightarrow \psi) \Rightarrow ((\psi \Leftrightarrow \phi) \Rightarrow (\varphi \Leftrightarrow \phi))$,
(s4) $(\varphi \Leftrightarrow \psi) \Rightarrow (\varphi \Rightarrow \psi)$,
(s5) $((\varphi \leftrightarrow \psi) \Leftrightarrow (\varphi \Rightarrow \varphi)) \Rightarrow ((\varphi \Leftrightarrow \phi) \leftrightarrow (\psi \Leftrightarrow \phi))$,
(s6) $\Box(\varphi \Leftrightarrow \psi) \Rightarrow (\Box \varphi \Leftrightarrow \Box \psi)$,
(s7) $\Diamond(\varphi \Leftrightarrow \psi) \Rightarrow (\Diamond \varphi \Leftrightarrow \Diamond \psi)$.

The deduction rules of **SMMTL** are:

- Modus Ponens (**MP**, φ and $\varphi \Rightarrow \psi$ infer ψ),
- Generalization (**G**, from φ infer $\Box\varphi$),
- Similarity (**S**, from φ, ψ infer $\varphi \Leftrightarrow \psi$).

The consequence relation $\vdash$ is define as follows, in the usual way. Let V be a theory (a set of formulas in **SMMTL**). A proof of a formula φ in V is a finite sequence of formulas with φ at its end, such that every formulas in the sequence is either an axiom of **SMMTL**, a formula of V, or the result of an application of an inference rule to previous formulas in the sequence. If a proof for φ exists in V, we say that can be deduced from V and we denote this by $V \vdash \varphi$.

Remark 5.1. *The proposition calculus* **SMMTL** *is an extension of* **MTL**. *Let V be a theory in* **MTL** *and φ be a formula of* **MTL** *such that $V \vdash_{\mathbf{MTL}} \varphi$. Then $V \vdash_{\mathbf{SMMTL}} \varphi$, since any V-proof in* **MTL** *is a V-proof in* **SMMTL**. *Conversely, $V_{\mathbf{MTL}} \subseteq V_{\mathbf{SMMTL}}$. Also, every theorem of* **MTL** *is a theorem of* **SMMTL**.

Definition 5.2. *Let $(L, \forall, \exists, S)$ be a similarity monadic MTL-algebra and V be a theory. An $(L, \forall, \exists, S)$-evaluation is a mapping e from the set of formulas of* **SMMTL** *to $(L, \forall, \exists, S)$ that satisfies, for each two formulas φ and ψ:*
  (1) $e(\varphi) \sqcap e(\psi) = e(\varphi) \cap e(\psi)$,
  (2) $e(\varphi) \sqcup e(\psi) = e(\varphi) \cup e(\psi)$,
  (3) $e(\varphi \& \psi) = e(\varphi) \otimes e(\phi)$,
  (4) $e(\varphi \Rightarrow \psi) = e(\varphi) \to e(\psi)$,
  (5) $e(\Box\varphi) = \forall e(\varphi)$,
  (6) $e(\Diamond\varphi) = \exists e(\varphi)$,
  (7) $e(\bar{0}) = 0$.

A $(L, \forall, \exists, S)$-evaluation e satisfies $e(\phi) = 1$, for every ϕ in V, it is called a $(L, \forall, \exists, S)$-model for V.

SMMTL is sound with respect to the variety of similarity monadic MTL-algebras, this is that if a formula φ can be deduced from a theory V in **SMMTL**, then for every similarity monadic MTL-algebra $(L, \forall, \exists, S)$ and for every $(L, \forall, \exists, S)$-model e of V, $e(\varphi) = 1$. Clearly, we need to verify the soundness of the new axioms and deduction rules of **SMMTL** (for the axioms and rules of **MTL**, the proofs (in **MTL**) can be copied). The fact that φ is a V-tautology with respect to a similarity monadic MTL-algebra $(L, \forall, \exists, S)$ will be denoted by $V \models_{(L,\forall,\exists,S)} \varphi$. If V is empty, a $\emptyset$-tautology with respect to a similarity monadic MTL-algebra $(L, \forall, \exists, S)$ will be simply called a tautology with respect to $(L, \forall, \exists, S)$ and the fact that φ is a tautology with respect to $(L, \forall, \exists, S)$ will be denoted by $\models_{(L,\forall,\exists,S)} \varphi$. And

$$\models_{(L,\forall,\exists,S)} \varphi \text{ if and only if } e(\varphi) = 1, \text{ for all } (L, \forall, \exists, S)\text{-evaluation}$$
$$e : V \longrightarrow (L, \forall, \exists, S).$$

Let φ and ψ be formulas, we define

$$\varphi \equiv_V \psi \text{ if and only if } V \vdash_{\mathbf{SMTL}} \varphi \Rightarrow \psi \text{ and } V \vdash_{\mathbf{MTL}} \psi \Rightarrow \varphi.$$

It is straightforward that the relation $\equiv_V$ is an equivalence relation. For any formula φ we will denote by $[\varphi]_V$ the equivalence class of φ with respect to $\equiv_V$. The set $Form_{\mathbf{SMMTL}}/\equiv_V = \{\,[\varphi]_V \mid \varphi \in Form_{\mathbf{SMMTL}}\,\}$ is the quotient of $Form_{\mathbf{SMMTL}}$ with respect to $\equiv_V$, where $Form_{\mathbf{SMMTL}}$ is the set of all formulas $\mathbf{SMMTL}$, we define the following operations:

$$[\varphi]_V \cap [\psi]_V := [\varphi \sqcap \psi]_V,$$
$$[\varphi]_V \cup [\psi]_V := [\varphi \sqcup \psi]_V,$$
$$[\varphi]_V \otimes [\psi]_V := [\varphi \& \psi]_V,$$
$$[\varphi]_V \to [\psi]_V := [\varphi \Rightarrow \psi]_V,$$
$$\forall [\varphi]_V := [\Box \varphi]_V,$$
$$\exists [\varphi]_V := [\Diamond \varphi]_V,$$
$$1_V := Theor_{\mathbf{SMMTL}}(V),$$
$$0_V := \neg 1_V,$$
$$S([\varphi]_V, [\psi]_V) := [\varphi \Leftrightarrow \psi]_V,$$

Proposition 5.3. *The following instances are realized from deduction rules of* SMMTL:
(a) $V \vdash (\varphi \Leftrightarrow \chi) \leftrightarrow (\psi \Leftrightarrow \chi)$ *if* $V \vdash \varphi \leftrightarrow \psi$,
(b) $V \vdash (\psi \Leftrightarrow \chi) \Rightarrow (\varphi \Rightarrow \chi)$ *if* $V \vdash \varphi \Rightarrow \psi$,
(c) $V \vdash (\chi \Leftrightarrow \varphi) \Rightarrow (\chi \Rightarrow \psi)$ *if* $V \vdash \varphi \Rightarrow \psi$,
(d) $V \vdash (\psi \Leftrightarrow \phi) \Rightarrow ((\varphi \Leftrightarrow \chi) \Rightarrow \phi)$ *if* $V \vdash \varphi$ *and* $V \vdash \chi \Rightarrow \psi$.

Proof. (a) Let $V \vdash (\varphi \leftrightarrow \psi)$. Then $V \vdash (\varphi \Rightarrow \psi)$. By **S**, we have $V \vdash (\varphi \Leftrightarrow \chi) \leftrightarrow (\psi \Rightarrow \chi)$, further by (s5) and **MP**, $V \vdash (\varphi \Leftrightarrow \chi) \leftrightarrow (\psi \Leftrightarrow \chi)$.
(b) If $V \vdash \varphi \Rightarrow \psi$, then by (s4), we have $\vdash (\psi \Leftrightarrow \chi) \Rightarrow (\psi \Rightarrow \chi)$. Then we have $\vdash (\psi \Rightarrow \chi) \Rightarrow ((\varphi \Rightarrow \psi) \Rightarrow (\varphi \Rightarrow \chi))$ and by the axioms of **MTL** and **MP**, $\vdash (\psi \Leftrightarrow \chi) \Rightarrow ((\varphi \Rightarrow \psi) \Rightarrow (\varphi \Rightarrow \chi))$. So $\vdash (\varphi \Rightarrow \psi) \Rightarrow ((\psi \Leftrightarrow \chi) \Rightarrow (\varphi \Rightarrow \chi))$. By hypothesis, $V \vdash (\psi \Leftrightarrow \chi) \Rightarrow (\varphi \Rightarrow \chi)$.
(c) Let $V \vdash \varphi \Rightarrow \psi$, then by (s4), we have $\vdash (\chi \Leftrightarrow \varphi) \Rightarrow (\chi \Rightarrow \varphi)$. Also $(\chi \Rightarrow \varphi) \Rightarrow ((\varphi \Rightarrow \psi) \Rightarrow (\chi \Rightarrow \psi))$, so by the axioms of **MTL** and **MP**, $\vdash (\chi \Leftrightarrow \varphi) \Rightarrow ((\varphi \Rightarrow \psi) \Rightarrow (\chi \Rightarrow \psi))$. So $\vdash (\varphi \Rightarrow \psi) \Rightarrow ((\chi \Leftrightarrow \varphi) \Rightarrow (\chi \Rightarrow \psi))$. By hypothesis, $V \vdash \varphi \Rightarrow \psi$. Thus $V \vdash (\chi \Leftrightarrow \varphi) \Rightarrow (\chi \Rightarrow \psi)$ by **MP**.
(d) Let $V \vdash \Rightarrow \psi$. Then by (c) we have $V \vdash (\varphi \Rightarrow \chi) \Rightarrow (\varphi \Rightarrow \psi)$. Also, $V \vdash \varphi \Rightarrow ((\varphi \Rightarrow \chi) \Rightarrow \psi)$. $V \vdash (\psi \Leftrightarrow \phi) \Rightarrow ((\varphi \Leftrightarrow \chi) \Rightarrow \phi)$ by (b). By hypothesis and **MP**, $V \vdash (\psi \Leftrightarrow \phi) \Rightarrow ((\varphi \Leftrightarrow \chi) \Rightarrow \phi)$. $\square$

Proposition 5.4. *The algebraic structure* $SMTL(V) = (Form_{\mathbf{SSMTL}}/\equiv_V, \cap, \cup, \otimes, \to, [0], [1], \forall, \exists, S)$ *is a similarity monadic MTL-algebra.*

Proof. By the above results, SMTL(V) is a monadic MTL-algebra, it follows that $(Form_{\mathbf{SSMTL}}/\equiv_V, \cap, \cup, \otimes, \to, [0], [1], \forall, \exists)$ is a monadic MTL-algebra. By Proposition 5.3 (a) the binary operation S is well defined. By Proposition 4.5, we have that S is a similarity on Form $Form_{\mathbf{SSMTL}}/\equiv_V$. □

Theorem 5.5. *(Soundness and completeness of **SMMTL**) A formula φ can be deduced from a thery V in **SMMTL** if and only if for every similarity monadic MTL-algebra $(L, \forall, \exists, S)$ and for every $(L, \forall, \exists, S)$-model e of V, $e(\varphi) = 1$.*

Proof. Let $\varphi_1, \varphi_2, ..., \varphi_n$ be a V-proof for φ in **SMMTL**, $(L, \forall, \exists, S)$ is a similarity monadic MTL- algebra and e an $(L, \forall, \exists, S)$-evaluation. We must prove that $e(\varphi_i) = 1$, for $i \in [n]$. Let φ_i be an axiom. It is clear that $e(\varphi_i) = 1$. If there are $j, k < i$ such that $\varphi_k = \varphi_j \to \varphi_i$. By hypothesis, $e(\varphi_k) = e(\varphi_j) = 1$, so $e(\varphi_i) = 1 \to e(\varphi_i) = e(\varphi_j \to \varphi_i) = e(\varphi_k) = 1$. If there are $j, k < i$ such that φ_i is $\varphi_j \Leftrightarrow \varphi_k$, then $e(\varphi_j) = e(\varphi_k) = 1$ so $e(\varphi_i) = S(e(\varphi_j), e(\varphi_k)) = S(1,1) = 1$. Thus $e(\varphi) = 1$, for $i = n$.

Conversely, by the fact that SMTL(V) is a a similarity monadic MTL-algebra, it is clear that $[\varphi]_V = 1$ in MTL(V). So φ deduced from V. □

Proposition 5.6. ***SMMTL** is conservative extensions of **MTL** this is if φ is an formula of **MTL** and V is a theory of **MTL**, then the following statements are equivalent:*
(1) $V \vdash_{\mathbf{MTL}} \varphi$,
(2) $V \vdash_{\mathbf{SSMTL}} \varphi$.

Proof. (1) $\Rightarrow$ (2) By Remark 5.1, the proof is clear.
(2) $\Rightarrow$ (1) Let $V \vdash_{\mathbf{SSMTL}}$ and $V \nvdash_{\mathbf{MTL}}$. Then there exists a linearly ordered monadic MTL-algebra $(L, \forall, \exists)$ and an $(L, \forall, \exists)$-evaluation $e : Form_{\mathbf{MTL}} \longrightarrow A$ such that $e(V) = 1$ and $e(\varphi) \neq 1$. It is clear that $S : (L, \forall, \exists) \times (L, \forall, \exists) \longrightarrow (L, \forall, \exists)$ define by $S(x, y) := x \leftrightarrow y$ is a similarity on $(L, \forall, \exists)$. So we consider $e_S : Form_{\mathbf{SSMTL}} \longrightarrow (L, \forall, \exists)$ to be the unique $(L, \forall, \exists)$-evaluation with $e_S(\varphi) = e(\varphi) \neq 1$, which is a contradiction, by Theorem 5.5. Thus $V \vdash_{\mathbf{MTL}} \varphi$. □

6 Conclusions

Motivated by the previous research of similarity MTL-algebras, we introduced and investigated similarity monadic MTL-algebras. We also studied similarity monadic filters and gave some characterizations of representable similarity monadic MTL-algebras. Finally, we introduced the logic of similarity monadic MTL-algebras and prove the completeness of them. Since the above topics are of current interest, we suggest further directions of research, focusing on the varieties of similarity monadic MTL-algebras. In particular, one can investigate semisimple, locally finite, finitely approximated and splitting varieties of similarity monadic MTL-algebras as well as varieties with the disjunction and the existence properties.

References

[1] F. G. Almiñana, M. E. Pelayes, Similarity DH-algebras, Journal of Algebraic Structures and Their Applications, **2**, (2015), 59-71.

[2] R. Belohlavek, Fuzzy relational systems. Foundations and Principles, Kluwer, (2002).

[3] R. A. Borzooei, S. K. Shoar, R. Ameri, Some types of filters in MTL-algebras, Fuzzy Sets and Systems, **187**, (2012), 92-102.

[4] D. Castaño, C. Cimadamore, J. P. D. Verela, L. Rueda, Monadic BL-algebras: The equivalent algebraic semantics of Hájek's monadic fuzzy logic, Fuzzy Sets and Systems, **320**, (2017), 40-59.

[5] J. L. Castro, F. Klawonn, Similarity in fuzzy reasoning using fuzzy logic, Mathware Soft Computation, **2**, (1995), 197-228.

[6] C. C. Chang, Algebraic analysis of many-valued logic, Transactions of The American Mathematical Society, **88**, (1958) 467-490.

[7] F. Chirteş, Similarity Łukasiewicz-Moisil algebras, Annals of the University of Cratova, Mathematics and Computer Science Series, **35**, (2008), 54-75.

[8] D. Dragulici, Quantifiers on BL-algebras, Analele Universitatii Bucuresti, **50**, (2001), 29-42.

[9] D. Dragulici, Polyadic BL-algebras. A representation theorem., J. Multiple-valued Logic and Soft Computing, 16, (2010), 265-302.

[10] D. Dubois, H. Prade, Similarity-based approximate reasoning, In: Computational intelligence Imitating Life, IEEE Press, New York, 1994, pp:69-80.

[11] F. Esteva, L. Godo, Monoidal t-norm based logic: towards a logic for left-continuoust-norms, Fuzzy Sets and Systems, **124**, (2001), 271-288.

[12] F. Formato, G. Gerla, M. Sessa, Similarity-based unification, Fundamenta Informaticae, **41**, (2000), 393-414.

[13] G. Georgescu, A. Popescu, Concept lattices and similarity in non-commutative fuzzy logic,Fundamenta Informaticae, **53**, (2002), 23-54.

[14] G. Georgescu, I. Leustean, A representation theorem for monadic Pavelka algebras, J. Universal Computers Science, 6, (2000), 105-111.

[15] I. Georgescu, Fuzzy Choice Functions, Springer, (2007).

[16] B. Gerla, I. Leuştean, Similarity MV-algebras, Fundamenta Informaticae, **69**, (2006), 287-300.

[17] S. Ghorbani, Monadic pseudo equality algebras, Soft Computing, **23**, (2019), 1499-1510.

[18] R. Grigolia, Monadic BL-algebras, Georgian Math. Journal, **13**, (2006), 267-276.

[19] P. Hájek, Metamathematics of Fuzzy logic, Kluwer Academic Publishers, Dordrech, 1998.

[20] P. Hájek, On fuzzy modal logics **S5(C)**, Fuzzy Sets and Systems, **161**, (2010), 2389–2396.

[21] R. P. Halmos, Algebraic logic, I. Monadic boolean algebras, Composition Mathematica, **12**, (1955), 217-249.

[22] J. Jacas, Similarity relations: The calculation of minimal generating families, Fuzzy Sets and Systems, **35**,(1990),151-161.

[23] S. Jenei, F. Montagna, A proof of standard completeness for Esteva and Godo's logic MTL, Studia Logica, **70**, (2002) 183-192.

[24] F. Klawonn, Similarity-based reasoning, In: Proceeding Third European Congress on Intelligent Techniques and Soft Computing, Aachem, 1995, pp:34-38.

[25] A. Di Nola, R. Grigolia, On monadic MV-algebras, Annals of Pure and Applied Logic, **128** (2004), 125-139.

[26] J. Rachůnek, D. Šalounová, Monadic bounded residuated lattices, Order, **30** (2013), 195-210.

[27] J. Rachůnek, F. Švrček, Monadic bounded residuated ℓ-monoids, Order, **25** (2008), 157-175.

[28] J. D. Rutledge, A preliminary investigation of the infinitely many-valued predicate calculus, Ph. D, Thesis, Cornell University, 1959.

[29] D. Scwartz, Polyadic MV-algebras, Math. Logic Quarterly, 26, (1980), 361-364.

[30] E. Turunen, A Łukasiewicz-style many-valued similarity reasoning, Review. in Beyond Two: Theory and Application of Multiple Valued Logic, Melvin Fitting and Ewa Orlowska editors, (2003), 311-321.

[31] J. T. Wang, A. Borumand Saeid, P. F. He, Similarity MTL-algebras and their corresponding logics, Journal of Multiple-Valued Logic and Soft Computing., **5**,(2019),607-628.

[32] J. T. Wang, X. L. Xin, P. F. He, Monadic bounded hoops, Soft Computing, **22**, (2018), 1749-1762.

[33] J. T. Wang, P. F. He, Y. H, She, Monadic NM-algebras, Logic Journal of the IGPL, **27**, (2019), 812-835.

[34] J. T. Wang, X. L. He, M. Wang. An algebraic study of the logic S5'(BL), Mathematica Slovaca, **72**, (2022), 1447-1462.

[35] J. T. Wang, X. L. Xin, Monadic algebras of an involutive monoidal t-norm based logic, Iranian Journal of Fuzzy Systems, 19(2022), 187-202.

[36] J. T. Wang, P. F. He, J. Yang, M. Wang, Monadic NM-algebras: an algebraic approach to monadic predicate nilpotent minimum logic, Journal of Logic and Computation, **32**, (2022), 741-766.

[37] J. T. Wang, Y. H. She, P. F. He, N. N. Ma, On Catrgorical equivalence of weak moandic residuated lattices and weak moandic c-differential residuated distributive lattices, Studia Logic, **111**, (2023): 361-390.

[38] J. T. Wang, H. W. Wu, P. F. He, Y. H. She, An algebraic proof of completeness for monadic fuzzy predicate logic MMTL∀, The Review of Symbolic Logic, (2023), DOI:https://doi.org/10.1017/S1755020323000291.

[39] J. T. Wang, M. Wang, Y. H. She, On algebraic semantics of similarity in monadic substructural predicate logics, Acta Electonica Sinica, **51**, (2023): 956-964.

[40] J. T. Wang, M. N. Kang, X. S. Fu, F. Li, State monadic residuated lattices and their corresponding filters, Journal of Intelligent and Fuzzy Systems, **44**, (2023):1739-1805.

[41] X. L. Xin, Y. L. Fu, Y. Y. Lai, J. T. Wang, Monadic pseudo BCI-algebras and corresponding logics, Soft Computing, **23**, (2019), 1499-1510.

[42] X. L. Xin, Y. J. Qin, P. F. He, Monadic operators on R_0-algebras, Fuzzy Systems and Mathematics, **30**, (2016), 48-57.

[43] L. Zadeh, Similarity relations and fuzzy orderings, Information Sciences, **2**, (1971),177-200.

[44] S. Zahiri, A. Borumand Saeid, Similarity monadic basic logic, Bulletin of the Belgian Mathematical Society Simon Stevin, **27**, (2020), 321-336.

Monadic operators on Bounded L-algebras

Lingling Mao
School of Mathematics, Northwest University, Xi'an 710127, P.R. China

Xiaolong Xin[*]
School of Mathematics, Northwest University, Xi'an 710127, P.R. China
`xlxin@nwu.edu.cn`

Xiaoguang Li
School of Science, Xi'an Aeronautical Institute, Xi'an 710077, P.R. China

Abstract

The main purpose of this paper is to investigate the type of monadic bounded L-algebras as L-algebras equipped with two monadic operators, named universal quantifier "∀" and existential quantifier "∃", respectively. First, we investigate the properties of pre-ideals on L-algebras and the pre-ideal generated by a nonempty subset of an L-algebra is defined. Second, we investigate monadic bounded L-algebras and monadic pre-ideals in monadic bounded L-algebras. Moreover, the relation between monadic bounded L-algebras and monadic quantum B-algebras is discussed. Finally, the relations among monadic self-similar L-algebras and other monadic structures are discussed, such as monadic (left) hoops, monadic Wajsberg hoops and monadic MV-algebras. Moreover, we obtain a characterization of monadic bounded L-algebras and monadic bounded self-similar L-algebras by relatively complete subalgebras and m-relatively complete subalgebras, respectively. These results are important to the further study of logical system with monadic operators.

Keywords: L-algebra; monadic operator; self-similarity; relatively complete algebra

[*]Corresponding author

1 Introduction

As an algebraic logic, the notion of L-algebras arose in the theory of one-sided lattice-ordered group and based upon the equation $(x \to y) \to (x \to z) = (y \to x) \to (y \to z)$ ([2, 11, 27, 15, 16]). On the other hand, L-algebra can be regarded as a solution of quantum Yang-Baxter equation ([17, 18]). Further, it was proved that for each L-algebra X there is a self-similar closure $S(X)$. At the same time, $S(X)$ admits a left hoop ([15]). For each L-algebra X, we say it has a negation if it has an element 0 such that there is a bijective mapping between x and x', where $x' = x \to 0$. Recently, it was also proved that each L-algebra that satisfies $x' \to y' = y \to x$ admits an MV-algebra ([31]), which indicates that L-algebras are generalizations of MV-algberas. Since L-algebras have been combined with quantum set ([21]), group theory ([22, 23, 24]), lattice theory ([25]) and other fields, the study of them have attracted more attention of many scholars.

In 1962, Halmos proposed Monadic Boolean algebras ([10]). It consists of Boolean algebras and the unary operation on it, where the unary operation is the algebraization of existential quantifier "∃". Moreover, the algebraization of existential quantifier "∃" and universal quantifier "∀" have also been studied in other non-classical logic ([13, 14]). From then on, monadic operators are introduced into more logic algebras such as monadic MV-algebras ([8, 9]), monadic Wajsberg hoops ([4]), monadic bounded hoops ([28]), monadic classes of quantum B-algebras ([6]), monadic involutive pseudo-BCK algebras ([12]) and monadic pseudo BCI-algebras ([32]) etc.([29, 30]). Accordingly, many good structures and properties have been obtained. As we all known, it has been turned out that L-algebras are structurally related to various algebras we mentioned above. We have known that although they are essentially different algebras, they are all particular types of L-algebras. As we said before, L-algebras are also logic algebras, so it is necessary to investigate their logical system. But the important work for monadic L-algebra is further to study the logical system with monadic operators. Above all, it is meaningful to extend universal quantifiers "∀" and existential quantifiers "∃" to L-algebras and to investigate the relationship between monadic L-algebras and other monadic algebras. This is the aim of our research on monadic L-algebras.

This paper is organized as follows: In Section 2, several fundamental definitions and properties of L-algebras used in this paper are recalled. In Section 3, the properties of pre-ideal on L-algebras and the pre-ideal generated by a nonempty subset in an L-algebra are investigated. In Section 4, we investigate monadic bounded L-algebra. Moreover, the relationship between monadic bounded L-algebras and monadic quantum B-algebras is discussed. In Section 5, we characterize monadic simple bounded L-algebras and study $MPI(L)$ from the point of their algebraic structures. In Section 6, the relations among monadic self-similar L-algebras and other monadic structures are discussed, such as monadic (left) hoops, monadic Wajsberg hoops and monadic MV-algebras. Moreover, we obtain a characterization of monadic bounded L-algebras and monadic bounded self-similar L-algebras by relatively complete subalgebras and m-relatively complete subalgebras, respectively.

2 Preliminaries

Some fundamental definitions and properties regarding L-algebras used in this paper are recalled in this section.

Definition 1. ([15]) *Given an algebra* $(L, \to, 1)$ *of type* $(2, 0)$, *we call it L-algebra if for any* $x, y, z \in L$, *the statements as follows are satisfied,*
(l_1) $1 \to x = x, x \to x = x \to 1 = 1,$
(l_2) $(y \to x) \to (y \to z) = (x \to y) \to (x \to z),$
(l_3) $x \to y = y \to x = 1$ *implies* $x = y$.

We call $(L, \to)$ a *cycloid* if it satisfies (l_2). From (l_1), we call 1 *logical unit* of L in terms of $\to$. Thus, the entailment relation is defined as $x \leq y \Leftrightarrow x \to y = 1$. Obviously, $\leq$ is a partial order relation on L.

Definition 2. ([15]) *Given an L-algebra* $(L, \to, 1)$, *we call it KL-algebra if for any* $x, y \in L$,
$$x \to (y \to x) = 1 \qquad (K).$$

Definition 3. ([15]) *Given an L-algebra* $(L, \to, 1)$, $I \subset L$ *is called an ideal if for any* $x, y \in L$, *the statements as follows are satisfied,*
(I_1) $1 \in I$,
(I_2) $x, x \to y \in I \Rightarrow y \in I$,
(I_3) $x \in I \Rightarrow (x \to y) \to y \in I$,
(I_4) $x \in I \Rightarrow y \to x, y \to (x \to y) \in I$.

From reference [7], we notice that if L satisfies $x \leq y \to x$, then the axiom (I_4) is redundant. If L satisfies $x \to (y \to z) = y \to (x \to z)$, then the axioms (I_3) and (I_4) are redundant.

Proposition 1. *([15]) Given an L-algebra* $(L, \to, 1)$. *Then for any* $x, y, z \in L$,
$$y \leq z \Rightarrow x \to y \leq x \to z,$$
which implies $x = y \Leftrightarrow x \to z = y \to z$. Furthermore,

L satisfies (K) $\Leftrightarrow x \leq y$ implies $y \to z \leq x \to z$.

Definition 4. ([5]) *Given an L-algebra* $(L, \to, 1)$, *we call it* CL-algebra *if it satisfies the exchange rule as follows, for all* $x, y, z \in L$,

$$x \to (y \to z) = y \to (x \to z) \qquad (exchange\ rule).$$

Remark 1. We notice that exchange rule implies (K) by taking $z := x$.

Definition 5. ([15]) *Given an L-algebra* $(X, \to, 1)$, *we call it* self-similar *if for any* $x \in X$, *there exists a bijection from downset* $\downarrow x := \{z \in X \mid z \leq x\}$ *onto X given by* $z \mapsto (x \to z)$.

We define a morphism $f : X \to Y$ between L-algebras X, Y to be a map which satisfies $f(1) = 1$ and $f(x \cdot y) = f(x) \cdot f(y)$ for all $x, y \in X$. If f is an inclusion $X \hookrightarrow Y$, we call X an L-subalgebra of Y. In case Y is a self-similar L-algebra with an L-subalgebra X which generates Y as a monoid, we call Y a self-similar closure of X.

Thus each $\downarrow x$ is in bijection with all of X. That is, there exists an inverse bijection $X \mapsto \downarrow x$ given by $y \mapsto y \cdot x$, which gives an everywhere defined multiplication on X that satisfies $x \to y \cdot x = y$, for all $x, y \in X$.

Proposition 2. ([26]) *A monoid* $(X, \cdot, \to, 1)$ *is a self-similar L-algebra iff for any* $x, y, z \in X$, *the statements as follows are satisfied,*
(1) $x \to y \cdot x = y$,
(2) $x \cdot y \to z = x \to (y \to z)$,
(3) $(x \to y) \cdot x = (y \to x) \cdot y$.

Statement (3) of Proposition 2 makes X into a $\wedge$-semilattice with $x \wedge y := (x \to y) \cdot x$.

Proposition 3. ([15]) *Given a KL-algebra X, then S(X) is commutative iff X satisfies*

$$(x \to y) \to y = (y \to x) \to x \qquad (C).$$

Definition 6. ([15]) *A* left hoop *is an algebra* $\langle H; \cdot, \to, 1 \rangle$ *s.t.* $\langle H; \cdot, 1 \rangle$ *is a monoid and the following statements hold, for any* $x, y, z \in H$,
(1) $x \to x = 1$,
(2) $x \cdot y \to z = x \to (y \to z)$,
(3) $(x \to y) \cdot x = (y \to x) \cdot y$.

If the binary operation $\cdot$ is commutative, then H is called *hoop*. A hoop H is called *Wajsberg hoop* if it satisfied (C)([4]). H is *bounded* if it has a bottom element

0 with respect to the order " $\leq$ ", which is defined by $x \leq y \Leftrightarrow x \to y = 1$. And the same as Wajsberg hoop.

Definition 7. ([3]) *Given an algebra $\langle A; \oplus, \odot, *, 0, 1 \rangle$ of type $(2,2,2,0,0)$, where $\langle A; \oplus, 0 \rangle$ is a commutative monoid, we call it an* MV-algebra *if the following statements hold, for any $x, y \in A$,*
(MV1) $x \oplus 1 = 1$,
(MV2) $x^{**} = x$,
(MV3) $0^* = 1$,
(MV4) $(x^* \oplus y)^* \oplus y = (x \oplus y^*)^* \oplus x$,
(MV5) $x \odot y = (x^* \oplus y^*)^*$.

Proposition 4. *([4]) Every Wajsberg algebra is equivalent to an MV-algebra.*

Definition 8. ([20]) *A* quantum B-algebra *is a partially ordered set $(X, \leq)$ with two binary operations $\to$ and $\rightsquigarrow$ satisfying the following axioms, for all $x, y, z \in X$,*
(QB1) $y \to z \leq (x \to y) \to (x \to z)$,
(QB2) $y \rightsquigarrow z \leq (x \rightsquigarrow y) \to (x \rightsquigarrow z)$,
(QB3) $y \leq z \Rightarrow x \to y \leq x \to z$,
(QB4) $x \leq y \to z \Leftrightarrow y \leq x \rightsquigarrow z$.

We denote it as $(X, \leq, \to, \rightsquigarrow)$. If for all $x, y \in X$, the equation $x \to y = x \rightsquigarrow y$ holds, we call $(X, \leq, \to, \rightsquigarrow)$ *commutative*. We call X *bounded* if it has a bottom element 0 and in such case, X also admits a largest element 1.

Definition 9. ([19, 20]) *A commutative quantum B-algebra $(X, \leq, \to)$ is called* integral *if there exists $u \in X$ such that the following hold for all $x \in X$:*
(1) $u \to x = x$,
(2) $x \to u = u$.

Proposition 5. *([5]) Every CL-algebra is a commutative integral quantum B-algebra.*

3 The pre-ideals of L-algebras

Definition 10. *Given an L-algebra $(L, \to, 1)$, $I \subset L$ is called a* pre-ideal *if for any $x, y \in L$, the statements as follows are satisfied:*
(PI_1) $1 \in I$,
(PI_2) $x, x \to y \in I \Rightarrow y \in I$.

We denote the collection of all pre-ideals of L as $PI(L)$.

Remark 2. (1) We notice that if $I, J \in PI(L)$, then $I \cap J \in PI(L)$. Indeed, if $x, x \to y \in I \cap J$, then $y \in I$ and $y \in J$, that is $y \in I \cap J$. Hence, $I \cap J \in PI(L)$. In such case, we say $I \wedge J = I \cap J$.

(2) From Definition 3, we note that if L satisfies exchange rule, then the definition of pre-ideal on L-algebras is the same as the definition of ideal on CL-algebras that introduced in reference [7].

Example 1. Given an algebra $L = \{x_1, x_2, 1\}$, where $x_1, x_2 \leq 1$, x_1 and x_2 are incomparable, the implication on L is defined as follows,

$\to$	x_1	x_2	1
x_1	1	x_2	1
x_2	x_1	1	1
1	x_1	x_2	1

Then $(L, \to, 1)$ is an L-algebra. We can check that all of the pre-ideals in L are $\{1\}$, $\{x_1, 1\}$, $\{x_2, 1\}$ and $\{x_1, x_2, 1\}$.

Example 2. Consider $L = [0, 1]$ and the implication on L is defined as $x \to y = 1$, if $x \leq y$; $x \to y = y$, others. Thus $(L, \to, 1)$ is an L-algebra. Put $I = (a, 1]$, where $a \geq 0$. Then one can check that I is a pre-ideal of L.

Definition 11. *Given an L-algebra $(L, \to, 1)$, a pre-ideal I is called* proper *if $I \neq L$. A proper pre-ideal I is said to be* prime, *if for any $I_1, I_2 \in PI(L)$ and $I_1 \cap I_2 \subseteq I$, then $I_1 \subseteq I$ or $I_2 \subseteq I$.*

Example 3. In Example 1, the pre-ideals $\{x_1, 1\}$ and $\{x_2, 1\}$ are prime.

Definition 12. *Given an L-algebra $(L, \to, 1)$, a proper pre-ideal is called* maximal *if it is not strictly contained in any other proper pre-ideal of L.*

Example 4. In Example 1, the maximal pre-ideals of L are $\{x_1, 1\}$ and $\{x_2, 1\}$.

Proposition 6. *Given an L-algebra $(L, \to, 1)$, $I \in PI(L)$. For any $x, y \in I, z \in L$, if $x \leq y \to z$, then $z \in I$.*

Proof. By Definition 10, it is immediate. □

Suppose that $\emptyset \neq X \subseteq L$, then the smallest pre-ideal of L which contains X, i.e. $\cap \{I \in PI(L) : X \subseteq I\}$ is said to be a pre-ideal of L generated by X, which is denoted by $\langle X \rangle$. To give a characterization of $\langle X \rangle$, we propose the following proposition.

Proposition 7. Given a KL-algebra $(L, \to, 1)$, then $x \to z \leq (y \to x) \to (y \to z)$ holds for all $x, y, z \in L$.

Proof. By (l_2) and (K), it is immediate. □

Theorem 1. Given a KL-algebra $(L, \to, 1)$, $\phi \neq X \subseteq L$. Then

$$<X> = \{a \in L \mid x_1 \to (x_2 \to (x_3 \to ...(x_n \to a)...)) = 1, \text{ for some } x_i \in X \text{ and } n \geq 1\}.$$

Proof. Put $M = \{a \in L \mid x_1 \to (x_2 \to (x_3 \to ...(x_n \to a)...)) = 1, \text{ for some } x_i \in X \text{ and } n \geq 1\}$. It is clear that $X \subseteq M$. Next we will show that M is a pre-ideal of L. Obviously, $1 \in M$. Now let $a, a \to b \in M$. Then there exist $x_1, x_2, ..., x_n, x_1', x_2', ..., x_m' \in X$, where $n, m \geq 1$, s.t.

$$x_1 \to (x_2 \to (x_3 \to ...(x_n \to a)...)) = 1$$

and

$$x_1' \to (x_2' \to (x_3' \to ...(x_m' \to (a \to b))...)) = 1.$$

Hence by Proposition 7, we have

$$a \to b \leq (x_n \to a) \to (x_n \to b) \leq (x_{n-1} \to (x_n \to a)) \to (x_{n-1} \to (x_n \to b)).$$

By repeating this way, we can get

$$a \to b \leq (x_1 \to (x_2 \to ...(x_n \to a)...)) \to (x_1 \to (x_2 \to ...(x_n \to b)...)).$$

Then by Proposition 1, we have

$$a \to b = 1 \to (x_1 \to (x_2 \to ...(x_n \to b)...)) \leq x_0 \to (x_1 \to (x_2 \to ...(x_n \to b)...)),$$

where $x_0 \in X$. Hence $x_m' \to (a \to b) \leq x_m' \to (x_1 \to (x_2 \to ...(x_n \to b)...))$.
Further, we can obtain
$x_1' \to (x_2' \to (x_m' \to (a \to b))...) \leq x_1' \to (x_2' \to ...(x_m' \to (x_0 \to (x_1 \to ...(x_n \to b)...)))...)$. Then

$$x_1' \to (x_2' \to ...(x_m' \to (x_0 \to (x_1 \to ...(x_n \to b)...)))...) = 1,$$

which implies $b \in M$. Therefore, $M \in PI(L)$. Let $I \in PI(L), X \subseteq I$ and $a \in M$. Then for some $x_i \in X$ and $n \geq 1$,

$$x_1 \to (x_2 \to (x_3 \to ...(x_n \to a)...))) = 1.$$

Since $1, x_1, x_2, ..., x_n \in I$, then $a \in I$. Therefore, M is the smallest pre-ideal that contains X, i.e. $M = \langle X \rangle$. □

Example 5. In Example 1, we can check that $\{x_1, x_2, 1\}$ is a KL-algebra. Take $X = \{x_1, x_2\} \subseteq X$, then Theorem 1 yields $\langle X \rangle = \{x_1, x_2, 1\}$.

In Theorem 1, we write $x \to (x \to ...(x \to a)...)$ as $x \stackrel{n}{\twoheadrightarrow} a$ and $x \stackrel{n}{\twoheadrightarrow} a = x \to (x \stackrel{n-1}{\twoheadrightarrow} a)$ for some $n \geq 1$. If $x \in L$ and $X = \{x\}$, we denote $\langle x \rangle$ as the pre-ideal generated by $\{x\}$ ($\langle x \rangle$ is said *principal*). The following corollary gives a characterization of principal pre-ideals.

Corollary 1. Given a KL-algebra $(L, \to, 1)$, for any $x \in L$,

$$\langle x \rangle = \{a \in L \mid x \stackrel{n}{\twoheadrightarrow} a = 1, for\ some\ n \geq 1\}.$$

Proof. Obviously, $b \in \{a \in L \mid x \stackrel{n}{\twoheadrightarrow} a = 1, for\ some\ n \geq 1\}$ implies $b \in \langle x \rangle$. Conversely, let $b \in \langle x \rangle$. Then by Theorem 1, for some $x_1 = x_2 = ... = x_m = x, m \geq 1$, s.t. $x \stackrel{m}{\twoheadrightarrow} b = 1$. Thus $b \in \{a \in L \mid x \stackrel{n}{\twoheadrightarrow} a = 1, for\ some\ n \geq 1\}$. Therefore, $\langle x \rangle = \{a \in L \mid x \stackrel{n}{\twoheadrightarrow} a = 1, for\ some\ n \geq 1\}$. □

In what follows, a characterization of a pre-ideal generated by two pre-ideals is obtained in CL-algebras.

Theorem 2. Given a CL-algebra $(L, \to, 1)$, $I_1, I_2 \in PI(L)$. Then

$$\langle I_1 \cup I_2 \rangle = \{a \in L \mid i \to (j \to a) = 1, for\ some\ i \in I_1, j \in I_2\}.$$

Proof. Obviously, $b \in \{a \in L \mid i \to (j \to a) = 1, for\ some\ i \in I_1, j \in I_2\}$ implies $b \in \langle I_1 \cup I_2 \rangle$. Conversely, let $b \in \langle I_1 \cup I_2 \rangle$. Then by Theorem 1, for some $i_1, i_2, \cdots, i_m \in I_1 (m \geq 1)$ and $j_1, j_2, \cdots, j_n \in I_2 (n \geq 1)$, s.t. $i_1 \to (i_2 \to \cdots (i_m \to (j_1 \to (j_2 \to \cdots (j_n \to b) \cdots))) \cdots) = 1$. Since $i_1, i_2, \cdots, i_m \in I_1 (m \geq 1)$ and $I_1 \in PI(L)$, we have $j_1 \to (j_2 \to \cdots (j_n \to b) \cdots) \in I_1$. So there exists $i \in I_1$ s.t. $i \to (j_1 \to (j_2 \to \cdots (j_n \to b) \cdots) \cdots) = 1$. By exchange rule, we obtain $j_1 \to (j_2 \to \cdots (j_n \to (i \to b) \cdots)) = 1$. Again since $I_2 \in PI(L)$, we can get $j \to (i \to b) = 1$ for some $i \in I_1, j \in I_2$. Therefore, $b \in \{a \in L \mid i \to (j \to a) = 1, for\ some\ i \in I_1, j \in I_2\}$. □

Corollary 2. Given a CL-algebra $(L, \to, 1)$, $x \in L$ and $I \in PI(L)$. Then

$$\langle I \cup \{x\} \rangle = \{a \in L \mid i \to (x \stackrel{n}{\twoheadrightarrow} a) = 1, for\ some\ i \in I\ and\ n \geq 1\}.$$

Proof. From Theorem 2, it is obvious. □

4 Monadic bounded L-algebras

The concept of monadic bounded L-algebra is given and several relevant properties of it are investigated in this section. Moreover, using these properties, we give a characterization of L-subalgebra. Finally, we discuss the relationship between monadic bounded L-algebras and monadic quantum B-algebras.

Definition 13. *We say an L-algebra* $(L, \rightarrow, 1)$ *is* bounded *if there exists an element* $0 \in L$, *s.t.* $0 \leq x$ *for all* $x \in L$.

Definition 14. *Given a bounded L-algebra* $(L, \rightarrow, 0, 1)$, *then* $(L, \rightarrow, 0, 1, \exists, \forall)$ $((L, \exists, \forall)$ *for short) of type* $(2, 0, 0, 1, 1)$ *is called a* monadic bounded L-algebra *(MBL-algebra for short) if for all* $x, y \in L$, *the axioms as follows are satisfied,*
(MBL1) $x \rightarrow \exists x = 1$,
(MBL2) $\forall x \rightarrow x = 1$,
(MBL3) $\forall (x \rightarrow \exists y) = \exists x \rightarrow \exists y$,
(MBL4) $\forall (\exists x \rightarrow y) = \exists x \rightarrow \forall y$,
(MBL5) $\exists \forall x = \forall x$.

In Definition 14, we call unary operators $\exists : L \longrightarrow L$ existential quantifier *and* $\forall : L \longrightarrow L$ universal quantifier, *respectively. Moreover, from (MBL1) and (MBL2) we can get that* $\exists$ *is an enlarge operation while* $\forall$ *is a reduce operation.*

Example 6. Let $L = \{0, x_1, x_2, x_3, 1\}$, where $0 \leq x_1 \leq x_2$, $x_3 \leq 1$, x_2 and x_3 are incomparable. The implication on L is defined as follows,

$\rightarrow$	0	x_1	x_2	x_3	1
0	1	1	1	1	1
x_1	0	1	1	1	1
x_2	0	x_3	1	x_3	1
x_3	0	x_2	x_2	1	1
1	0	x_1	x_2	x_3	1

Then $(L, \rightarrow, 0, 1)$ is a bounded L-algebra. Define $\exists$ and $\forall$ as $\exists 0 = 0$, $\exists x_1 = \exists x_3 = x_3$, $\exists 1 = \exists x_2 = 1$; $\forall 0 = \forall x_1 = \forall x_2 = 0$, $\forall x_3 = x_3$, $\forall 1 = 1$. Then the axioms (MBL1)-(MBL5) are satisfied, which implies $(L, \exists, \forall)$ is an MBL-algebra.

Example 7. Consider the bounded L-algebra in Example 2. For any $x \in L$, we define $\forall$ and $\exists$ as follows,

$$\forall x = \begin{cases} 1, & x = 1 \\ 0, & x \neq 1 \end{cases}, \quad \exists x = \begin{cases} 0, & x = 0 \\ 1, & x \neq 0 \end{cases}.$$

Then the axioms (MBL1)-(MBL5) are satisfied, which implies $(L, \exists, \forall)$ is an MBL-algebra.

Now, we give some properties of operators $\exists$ and $\forall$ on monadic bounded L-algebras.

Proposition 8. Given an MBL-algebra $(L, \exists, \forall)$, then for any $x, y \in L$, the statements as follows are satisfied,
(1) $\forall 0 = \exists 0 = 0, \forall 1 = \exists 1 = 1$,
(2) $\forall \exists x = \exists x$,
(3) $\exists \exists x = \exists x$,
(4) $\forall \forall x = \forall x$,
(5) $\forall (\exists x \to \exists y) = \exists x \to \exists y$,
(6) $\exists (\exists x \to y) \leq \exists x \to \exists y$,
(7) $\forall (\forall x \to y) = \forall x \to \forall y$,
(8) $\forall x = x \Leftrightarrow \exists x = x$,
(9) $\forall (\forall x \to \exists y) = \forall x \to \exists y$,
(10) $\forall (x \to \forall y) = \exists x \to \forall y$,
(11) $\forall (\forall x \to \forall y) = \forall x \to \forall y$,
(12) $\exists (\exists x \to \exists y) = \exists x \to \exists y$,
(13) $\exists (\forall x \to \forall y) = \forall x \to \forall y$,
(14) $x \leq y \Rightarrow \forall x \leq \forall y, \exists x \leq \exists y$,
(15) $x \leq \exists y \Leftrightarrow \exists x \leq \exists y, \forall x \leq y \Leftrightarrow \forall x \leq \forall y$,
(16) if L satisfies (K), then $\forall (x \to y) \leq \forall x \to \forall y$,
(17) $\forall (x \to y) \leq \exists x \to \exists y$,
(18) if L satisfies (K), then $\forall ((x \to \forall y) \to \forall y) = (\forall x \to \forall y) \to \forall y$,
(19) if L satisfies (K), then $\forall ((x \to \forall y) \to x) = (\forall x \to \forall y) \to \forall x$.

Proof. In here, the proof of statements (1)-(17) are similar to the proof of Proposition 3.5 in ([28]), so we just give the proof of statements (18) and (19).
(18) Assume the L-algebra satisfies (K). On the one hand, by (10), Proposition 1 and (13), we have $\forall ((x \to \forall y) \to \forall y) = \exists (x \to \forall y) \to \forall y \geq \exists (\forall x \to \forall y) \to \forall y = (\forall x \to \forall y) \to \forall y$. On the other hand, by similar way, we have $\forall ((x \to \forall y) \to \forall y) = \exists (x \to \forall y) \to \forall y \leq \forall (x \to \forall y) \to \forall y \leq (\forall x \to \forall \forall y) \to \forall y = (\forall x \to \forall y) \to \forall y$. Hence, $\forall ((x \to \forall y) \to \forall y) = (\forall x \to \forall y) \to \forall y$.
(19) It is verified by taking $x := \forall y$ in statement (18). □ □

In any bounded L-algebra L, 0 is the smallest element, then for any $x \in L$, it is convenient to denote $x' = x \to 0$.

Proposition 9. Given an MBL-algebra $(L, \exists, \forall)$, then for any $x, y \in L$, the statements as follows are satisfied,
(1) $\forall x' \leq (\forall x)'$,
(2) $(\forall x)' = \forall(\forall x)'$,
(3) $(\exists x)' = \exists(\exists x)' = \forall x'$,
(4) $(\exists x')' = \forall x$.

Proof. (1) It is immediate by Proposition 8(16).

(2) By Proposition 8(1), (MBL5) and (MBL3), $(\forall x)' = \forall x \to \exists 0 = \exists \forall x \to \exists 0 = \forall(\forall x \to 0) = \forall(\forall x)'$.

(3) By Proposition 8(1) and (MBL3), $(\exists x)' = \exists x \to 0 = \exists x \to \exists 0 = \forall(x \to \exists 0) = \forall(x \to 0) = \forall(x')$. Further, by Proposition 8(1), 85(3), (MBL3) and (MBL5), we have $\exists(\exists x)' = \exists(\exists x \to \exists\exists 0) = \exists(\forall(x \to 0)) = \exists \forall x' = \forall x'$. Therefore, $(\exists x)' = \exists(\exists x)' = \forall(x')$.

(4) By (MBL3) and Proposition 8(1), it is immediate. □ □

Definition 15. ([1]) *Let $f : A \to B$ and $g : B \to A$ are two order-preserving mappings, where A and B are posets. We call the pair (f, g) a Galois connection between A and B if $fg \geq id_A$ and $gf \leq id_B$.*

Proposition 10. Given an MBL-algebra $(L, \exists, \forall)$, then the pair $(\exists, \forall)$ establishes a Galois connection over $(L, \leq)$.

Proof. By Proposition 8(2) and (MBL5), $\forall \exists x = \exists x \geq x = id_L(x)$ and $\exists \forall x = \forall x \leq x = id_L(x)$ for all $x \in L$. Hence, $(\exists, \forall)$ establishes a Galois connection over L. □ □

For each MBL-algebra $(L, \exists, \forall)$, we denote
$$L_{\exists\forall} = \{x \in L | \exists x = x\} = \{x \in L | \forall x = x\}.$$

Example 8. It is obvious that $L_{\exists\forall} = \{0, x_3, 1\}$ in Example 6.

The following proposition gives a characterization of L-subalgebra by $L_{\exists\forall}$ and some properties of $L_{\exists\forall}$.

Proposition 11. Given an MBL-algebra $(L, \exists, \forall)$. Then the following statements hold,
(1) $L_{\exists\forall}$ is a subalgebra of $(L, \to, 0, 1)$,
(2) $\forall L = L_{\exists\forall} = \exists L$,
(3) If $L_{\exists_i \forall_i} = L_{\exists_j \forall_j}$, then $\exists_i = \exists_j$ and $\forall_i = \forall_j$ $(i \neq j)$,
(4) If $Im(\forall_i) = Im(\forall_j)$, then $\forall_i = \forall_j$ $(i \neq j)$,
(5) If $Im(\exists_i) = Im(\exists_j)$, then $\exists_i = \exists_j$ $(i \neq j)$.

Proof. (1) By Proposition 8(1), we know $0, 1 \in L_{\exists\forall}$ and so $L_{\exists\forall} \neq \emptyset$. Now we only need to prove that the operation $\to$ is closed for $L_{\exists\forall}$. Assume $x, y \in L_{\exists\forall}$. By (MBL3), $\forall(x \to y) = \forall(x \to \exists y) = \exists x \to \exists y = x \to y$, which means $x \to y \in L_{\exists\forall}$.

(2) Assume $y \in \forall L$, then there exists $x \in L$ s.t. $y = \forall x$. From Proposition 8(4), $\forall y = \forall\forall x = \forall x = y$, that is $y \in L_{\exists\forall}$. Conversely, if $y \in L_{\exists\forall}$, then $y = \forall y \in \forall L$. Therefore, $L_{\exists\forall} = \forall L$. Similarly, $\exists L = L_{\exists\forall}$.

(3) By Proposition 8(4), $\forall_i \forall_i x = \forall_i x$, so $\forall_i x \in L_{\exists_i \forall_i} = L_{\exists_j \forall_j}$, hence $\forall_j \forall_i x = \forall_i x$ for all $x \in L$, which means $\forall_j \forall_i = \forall_i$. Similarly, $\forall_i \forall_j = \forall_j$. Further, by (MBL2) and Proposition 8(14), $\forall_i x = \forall_j \forall_i x \leq \forall_j x$ and $\forall_j x = \forall_i \forall_j x \leq \forall_i x$ for each $x \in L$. Hence, $\forall_i = \forall_j$, then we have $\exists_i = \exists_j$ by Proposition 8(8).

By statements (1) and (2), the statements (4) and (5) are immediate. □ □

In what follows, we discuss the relationship between MBL-algebras and monadic quantum B-algebras.

Definition 16. ([6]) *Given a quantum B-algebra* $(X, \leq, \to, \leadsto, u, \exists, \forall)$ *with unital element* u, *we call* $(X, \exists, \forall)$ *a monadic quantum B-algebra, if the following axioms are satisfied, for any* $x, y \in X$,
(MQB1) $x \leq \exists x$,
(MQB2) $\forall x \leq x$,
(MQB3) $\forall(x \to \exists y) = \exists x \to \exists y$, $\forall(x \leadsto \exists y) = \exists x \leadsto \exists y$,
(MQB4) $\forall(\exists x \to y) = \exists x \to \forall y$, $\forall(\exists x \leadsto y) = \exists x \leadsto \forall y$,
(MQB5) $\exists\forall x = \forall x$,
(MQB6) $\forall u = \exists u = u$.

If X is commutative and integral, then we call $(X, \exists, \forall)$ a *monadic commutative integral quantum B-algebra*.

Theorem 3. *Every MBL-algebra that satisfies exchange rule is a monadic commutative integral quantum B-algebra.*

Proof. Let $(X, \exists, \forall)$ be a MBL-algebra that satisfies exchange rule, then from Proposition 5, it is a commutative integral quantum B-algebra. Moreover, axioms (MQB1)-(MQB6) are directly obtained from (MBL1) to (MBL5) and Proposition 8(1), respectively. Therefore, $(X, \exists, \forall)$ is a monadic commutative integral quantum B-algebra. □ □

5 Monadic pre-ideals in monadic bounded L-algebras

In this section, we study monadic pre-ideal, maximal monadic pre-ideal and prime monadic pre-ideal in MBL-algebras. Moreover, some characterizations of them are obtained from the point of universal quantifiers and existential quantifiers, respectively. In addition, it is shown that the set of all monadic pre-ideals of MBL-algebras is a lattice under the inclusion order $\subseteq$ and an equivalent characterization of it is obtained.

Definition 17. *Given an MBL-algebra $(L, \exists, \forall)$ and $I \in PI(L)$. We call I a monadic pre-ideal if for all $x \in I$, we have $\forall x \in I$.*

In Definition 17, we call I a *maximal monadic pre-ideal* if I is the maximal pre-ideal of L. The collection of all monadic pre-ideals of $(L, \exists, \forall)$ is denoted by $MPI(L)$.

Example 9. Given an MBL-algebra $(L, \exists, \forall)$. Then $Ker(\forall) = \{x \in L | \forall x = 1\} \in MPI(L)$.

Example 10. Consider the Example 6, it is straightforward to verify that all of the monadic pre-ideals in $(L, \exists, \forall)$ are $\{1\}$, $\{x_3, 1\}$ and $\{0, x_1, x_2, x_3, 1\}$. Furthermore, $\{x_1, x_2, x_3, 1\} \in PI(L)$ but $\{x_1, x_2, x_3, 1\} \notin MPI(L)$ since $\forall x_1 = \forall x_2 = 0 \notin \{x_1, x_2, x_3, 1\}$.

Given an MBL-algebra $(L, \exists, \forall)$ and $\emptyset \neq X \subseteq L$, we denote $\langle X \rangle_\forall$ as monadic pre-ideal of L generated by X, i.e. $\langle X \rangle_\forall$ is the smallest monadic pre-ideal of $(L, \exists, \forall)$ containing X. Now we give a characterization of $\langle X \rangle_\forall$ in the following proposition.

Proposition 12. *Given an MBL-algebra $(L, \exists, \forall)$ satisfies (K), $\emptyset \neq X \subseteq L$. Then*

$$\langle X \rangle_\forall = \{a \in L \mid \forall x_1 \to (\forall x_2 \to (\forall x_3 \to \cdots (\forall x_n \to a) \cdots)) = 1, \, for \, some \, x_i \in X, n \geq 1\}.$$

Proof. The proof for it is similar to that of Theorem 1. □ □

Proposition 13. *Given an MBL-algebra $(L, \exists, \forall)$, $I, I_1, I_2 \in MPI(L)$ and $x \notin I$. The following statements hold,*
(1) if L satisfies (K), then $\langle x \rangle_\forall = \{a \in L \mid \forall x \xrightarrow{n} a = 1, \, for \, some \, n \geq 1\}$,
(2) if L satisfies exchange rule, then $\langle I_1 \cup I_2 \rangle_\forall = \{a \in L \mid \forall i \to (\forall j \to a) = 1, \, for \, some \, i \in I_1, j \in I_2 \, and \, n \geq 1\}$,
(3) if L satisfies exchange rule, then $\langle I \cup \{x\} \rangle_\forall = \{a \in L \mid \forall i \to (\forall x \xrightarrow{n} a) = 1, \, for \, some \, i \in I \, and \, n \geq 1\}$.

Proof. From Corollary 1, Theorem 2, Corollary 2 and Proposition 12, they are immediate. □ □

Remark 3. From Remark 1, Remark 2(2), we note that the conclusions in Theorem 1, Corollary 1, Proposition 12 and Proposition 13(1) are also hold on CL-algebras that introduced in reference [7].

Corollary 3. Given an MBL-algebra $(L, \exists, \forall)$ satisfies (K). Then for any $x, y \in L$, the statements as follows are satisfied,
(1) $\langle \forall x \rangle_\forall = \langle x \rangle_\forall$,
(2) if $x \leq y$, then $\langle y \rangle_\forall \subseteq \langle x \rangle_\forall$.

Proof. (1) By Propositions 8(4) and 13(1), it is immediate.
(2) By Propositions 8(14) and 13(1), it is immediate. □ □

Proposition 14. Given an MBL-algebra $(L, \exists, \forall)$ that satisfies exchange rule, $x \in L_{\exists \forall}$ and $I \in MPI(L)$, then $\langle I \cup \{x\} \rangle \in MPI(L)$.

Proof. Assume that $a \in \langle \{I \cup \{x\}\} \rangle$, then by Corollary 2, there exist some $i \in I$ and $n \geq 1$ s.t. $i \to (x \stackrel{n}{\twoheadrightarrow} a) = 1$. Since $i \in I$ and $I \in MPI(L)$, $x \stackrel{n}{\twoheadrightarrow} a \in I$ and further, $\forall (x \stackrel{n}{\twoheadrightarrow} a) \in I$. Then there exists $j \in I$ s.t. $\forall (x \stackrel{n}{\twoheadrightarrow} a) = j$. Hence $j \to \forall (x \stackrel{n}{\twoheadrightarrow} a) = 1$. Since $x \in L_{\exists \forall}$, then by Proposition 8(16) and Proposition 1, $\forall (x \stackrel{n}{\twoheadrightarrow} a) \leq \forall x \stackrel{n}{\twoheadrightarrow} \forall a = x \stackrel{n}{\twoheadrightarrow} \forall a$. So $1 = j \to \forall (x \stackrel{n}{\twoheadrightarrow} a) \leq j \to (x \stackrel{n}{\twoheadrightarrow} \forall a)$, which implies $j \to (x \stackrel{n}{\twoheadrightarrow} \forall a) = 1$, hence $\forall a \in \langle I \cup \{x\} \rangle$. Therefore, $\langle I \cup \{x\} \rangle \in MPI(L)$. □ □

Inspired by Proposition 14, the following proposition gives an equivalent characterization of maximal monadic pre-ideal from the point of universal quantifiers and existential quantifiers, respectively.

Proposition 15. Given an MBL-algebra $(L, \exists, \forall)$ that satisfies exchange rule, $I \in MPI(L)$ and I is proper. Then the following statements are equivalent,
(1) I is maximal,
(2) for each $x \in L$, $\forall x \in I$ or $(\forall x)' \in I$,
(3) for each $x \in I$, $\exists x \in I$ or $(\exists x)' \in I$.

Proof. (1) $\Rightarrow$ (2) Let I is maximal and there exists $x \in L$ s.t. $\forall x, (\forall x)' \notin I$. Consider the pre-ideal $\langle \{\forall x\} \cup I \rangle$, it is proper since $(\forall x)' \notin I$. Moreover, from Proposition 14, $\langle \{\forall x\} \cup I \rangle \in MPI(L)$, so $I \subset \langle \{\forall x\} \cup I \rangle$, which conflicts with the fact that I is maximal.

(2) $\Rightarrow$ (1) Suppose for any $x \in L, \forall x \in I$ or $(\forall x)' \in I$, but I is not maximal. Then there is a proper monadic pre-ideal D s.t. $I \subset D$, i.e., there exists $x \in D$ but $x \notin I$. Therefore, $\forall x \notin I$ but $(\forall x)' \in I$, so $(\forall x)' \in D$. Moreover, since $x \in D$ and

$D \in MPI(L)$, then $\forall x \in D$, which conflicts with the fact that $(\forall x)' \in D$.
(1) $\Leftrightarrow$ (3) The proof is similar to the proof of (1) $\Leftrightarrow$ (2). □ □

The following theorem gives an equivalent characterization of monadic pre-ideals.

Theorem 4. ([7]) Given an MBL-algebra $(L, \exists, \forall)$ satisfies exchange rule, $I \in PI(L)$. Then we have equivalent statements as follows,
(1) $I \in MPI(L)$,
(2) $I = \langle I \cap L_{\exists\forall} \rangle$.

Now we introduce monadic simple bounded L-algebras and some equivalent characterizations of them are given.

Definition 18. *An MBL-algebra $(L, \exists, \forall)$ is called* simple *if there are only two monadic pre-ideals, which are L and $\{1\}$.*

Theorem 5. Given an MBL-algebra $(L, \exists, \forall)$ that satisfies (K). Then we have equivalent statements as follows,
(1) $(L, \exists, \forall)$ is simple,
(2) $\forall L$ is simple,
(3) $\exists L$ is simple,
(4) $L_{\exists\forall} = \{0, 1\}$.

Proof. (1) $\Rightarrow$ (2) Suppose $(L, \exists, \forall)$ is simple, $I \in MPI(\forall L)$ and $I \neq \{1\}$. Now we only need to prove that $\forall L = I$. Consider the set $I_i = \{a \in L \mid i \to a = 1,$ *for a certain* $i \in I\}$. It is clear that $1 \in I_i$. Now assume $x, x \to y \in I_i$, then there exist $i_1, i_2 \in I$ s.t. $i_1 \to x = 1, i_2 \to (x \to y) = 1$. So $i_1 \leq x$, then $x \to y \leq i_1 \to y$. Further, from $1 = i_2 \to (x \to y) \leq i_2 \to (i_1 \to y)$, we have $i_2 \to (i_1 \to y) = 1$ and hence $y \in I$ since $I \in MPI(\forall L)$. Then there exists a certain $i \in I$ s.t. $y = i$, we have $i \to y = 1$, which implies $y \in I_i$. Therefore, $I_i \in PI(L)$. Moreover, if $x \in I_i$, then there exists $i \in I$ s.t. $i \to x = 1$. By Proposition 8(16), $1 = \forall(i \to x) \leq \forall i \to \forall x = i \to \forall x$ since $i \in \forall L$. So $i \to \forall x = 1$, which implies $\forall x \in I_i$. Therefore, $I_i \in MPI(L)$. Since $(L, \forall, \exists)$ is simple and $I \subseteq I_i$, $I_i \neq \{1\}$, we have $I_i = L$. So $0 \in I_i$, hence $I = \forall L$, which means $\forall L$ is simple.
(2) $\Rightarrow$ (1) Let $I \in MPI(L)$, then $I \cap \forall L \in MPI(\forall L)$, hence $I \cap \forall L = \{1\}$ or $I \cap \forall L = \forall L$.
(i) If $I \cap \forall L = \forall L$, then $\forall L \subseteq I$ and, since $0 \in \forall L$, we can get that $I = L$.
(ii) If $I \cap \forall L = \{1\}$ and $x \in I$, then $\forall x \in I \cap \forall L$, so $\forall x = 1$, that is, $x = 1$(if $x \neq 1$, which conflicts with $\forall L$ is simple), so $I = \{1\}$. Then we conclude that $(L, \forall, \exists)$ is simple.
(2) $\Leftrightarrow$ (3) Since $\forall L = \exists L$ by Proposition 11, it is immediate.

(2) $\Leftrightarrow$ (4) Since $\forall L = L_{\exists\forall}$ by Proposition 11, it is immediate. □ □

Theorem 5 offers a way to check whether an MBL-algebra is simple. As an application, we can use it to verify that the MBL-algebra in Example 7 is simple since $L_{\exists\forall} = \{0, 1\}$.

In what follows, the concept of prime monadic pre-ideal of MBL-algebras is presented. Moreover, we get that each maximal monadic pre-ideal is prime.

Definition 19. *Given an MBL-algebra $(L, \forall, \exists)$, $I \in MPI(L)$ and $I \neq L$, we call I prime monadic pre-ideal, if for all $I_1, I_2 \in MPI(L)$ s.t. $I_1 \cap I_2 \subseteq I$, then $I_1 \subseteq I$ or $I_2 \subseteq I$.*

Example 11. In Example 10, we can check that $\{x_3, 1\}$ is a prime monadic pre-ideal of $(L, \forall, \exists)$.

Proposition 16. *Given an MBL-algebra $(L, \exists, \forall)$. $I \in MPI(L)$ and I is maximal, then I is prime.*

Proof. Let $I_1, I_2 \in MPI(L)$ with $I_1 \cap I_2 \subseteq I$. If $I_1 \not\subseteq I$, then $I_1 = L$ since I is a maximal monadic pre-ideal, it follows that $I_1 \cap I_2 = L \cap I_2 = I_2 \subseteq I$. Therefore, I is prime. □ □

In what follows, we prove that the collection of all monadic pre-ideals of an L-algebra is a lattice. Further, the isomorphic relation between lattice $(MPI(L), \wedge, \vee)$ and lattice $PI(L_{\exists\forall})$ is discussed.

Proposition 17. *Given an MBL-algebra $(L, \exists, \forall)$ that satisfies exchange rule. For any $I_1, I_2 \in MPI(L)$, we define $I_1 \wedge I_2 = I_1 \cap I_2$, $I_1 \vee I_2 = \langle I_1 \cup I_2 \rangle_\forall$. Then $(MPI(L), \wedge, \vee)$ is a lattice under the inclusion order $\subseteq$.*

Proof. Assume that $\{I_j\}_{j \in K}$ is a family of monadic pre-ideals of $(L, \exists, \forall)$. Obviously, according to Proposition 13(2), the infimum is $\wedge_{j \in K} I_j = \cap_{j \in K} I_j$ and the supremum is $\vee_{i \in K} I_i = \{a \in L \mid \forall i_{j_1} \to (\forall i_{j_2} \to \cdots (\forall i_{j_m} \to a) \cdots) = 1, \text{ for some } i_{j_k} \in I_{j_k}, j_k \in K\}$. Therefore, $(MPI(L), \wedge, \vee)$ is a lattice under the inclusion order $\subseteq$. □ □

Theorem 6. *Given an MBL-algebra $(L, \exists, \forall)$ that satisfies (K). Then $(MPI(L), \wedge, \vee)$ is isomorphic to the lattice $PI(L_{\exists\forall})$ (that is, the set of all pre-ideals of bounded L-algebra $L_{\exists\forall}$).*

Proof. Consider the mappings $\varphi : MPI(L) \longrightarrow PI(L_{\exists\forall})$ given by $\varphi(I) = I \cap L_{\exists\forall}$, for any $I \in MPI(L)$, and $\psi : PI(L) \longrightarrow MPI(L)$ given by $\psi(J) = \langle J \rangle_\forall$, for any $J \in PI(L)$. Then we can check that φ and ψ are well-defined. Next, for any $I \in MPI(L)$ and $J \in PI(L_{\exists\forall})$, from the definition of φ and ψ, it is easy to verify that $\varphi(\psi(J)) = J$ and $\psi(\varphi(I)) = I$. Moreover, if $J_1, J_2 \in PI(L_{\exists\forall})$ s.t. $J_1 \subseteq J_2$, then $\langle J_1 \rangle_\forall \subseteq \langle J_2 \rangle_\forall$, which means $\psi(J_1) \subseteq \psi(J_2)$. On the other hand, if $I_1, I_2 \in MPI(L)$ s.t. $I_1 \subseteq I_2$, then $I_1 \cap L_{\exists\forall} \subseteq I_2 \cap L_{\exists\forall}$, which means $\varphi(I_1) \subseteq \varphi(I_2)$. Then we conclude that $(MPI(L), \wedge, \vee)$ is isomorphic to $PI(L_{\exists\forall})$. □

6 Relations between monadic bounded self-similar L-algebras and other related monadic algebras

Relations between monadic bounded self-similar L-algebras and other monadic structures are discussed in this section, such as monadic (left) hoops, monadic Wajsberg hoops and monadic MV-algebras. Moreover, we obtain a characterization of monadic bounded L-algebras and monadic bounded self-similar L-algebras by relatively complete subalgebras and m-relatively complete subalgebras, respectively.

In order to investigate monadic bounded self-similar L-algebras, we give the following propositions of self-similar L-algebras.

Proposition 18. Given a self-similar L-algebra $(X, \to, \cdot, 1)$. Then for any $x, y, z \in X$, the statements as follows are satisfied,
(1) $x \cdot y \leq z \Leftrightarrow x \leq y \to z$,
(2) $(x \to y) \cdot x = (y \to x) \cdot y \leq y$,
(3) if $x \leq y$, then $x \cdot z \leq y \cdot z$,
(4) if X satisfies (K) and $x \leq y$, then $z \cdot x \leq z \cdot y$,
(5) if X satisfies (K), then $(y \to z) \cdot (x \to y) \leq x \to z$,
(6) $x \to y = x \cdot z \to y \cdot z$.

Proof. The statements (1), (2), (3) and (6) are directly obtained from Proposition 2(2).
(4) Assume X satisfies (K) and $x \leq y$, then by Proposition 1, we have $z = y \to z \cdot y \leq x \to z \cdot y$. Hence, $z \cdot x \leq z \cdot y$.
(5) Assume X satisfies (K), then by Proposition 7, it is valid. □

Proposition 19. Given a bounded self-similar L-algebra $(X, \to, \cdot, 0, 1)$. Then for any $x \in X$, the statements as follows are satisfied,
(1) $x \cdot 1 = 1 \cdot x = x$,
(2) $x \cdot 0 = 0 \cdot x = 0$.

Proof. (1) By Proposition 2(2), $x \cdot 1 \to x = x \to (1 \to x) = x \to x = 1$, that is $x \cdot 1 \leq x$. Moreover, by Proposition 18(1), $x \leq 1 \to x \cdot 1 = x \cdot 1$. Hence, $x \cdot 1 = x$.

Similarly, $1 \cdot x = x$.
(2) Since $x \leq 1 = 0 \rightarrow 0$, we have $x \cdot 0 \leq 0$ and thus $x \cdot 0 = 0$. Similarly, $0 \cdot x = 0$.
□ □

Definition 20. *Given a bounded self-similar L-algebra* $(X, \rightarrow, \cdot, 0, 1)$. *Then we call* $(X, \rightarrow, \cdot, 0, 1, \exists, \forall)$ *(*$(X, \exists, \forall)$ *for short) of type (2,2,0,0,1,1) a monadic bounded self-similar L-algebra (MBSL-algebra for short) if for any* $x, y \in X$, *it satisfies axioms (MBL1)-(MBL5).*
Moreover, we call an MBSL-algebra strong if it satisfies the axiom
(MBL6) $\exists(x \cdot x) = \exists x \cdot \exists x$.

Proposition 20. Given an MBSL-algebra $(X, \exists, \forall)$. Then for any $x, y \in X$, the statements as follows are satisfied,
(1) $\exists(\exists x \cdot \exists y) = \exists x \cdot \exists y$,
(2) $\forall(\exists x \cdot \exists y) = \exists x \cdot \exists y$,
(3) $\forall(\forall x \cdot \forall y) = \forall x \cdot \forall y$,
(4) $\exists(\forall x \cdot \forall y) = \forall x \cdot \forall y$,
(5) $\exists(\exists x \wedge \exists y) = \exists x \wedge \exists y$,
(6) $\forall(x \wedge y) = \forall x \wedge \forall y$.

Proof. (1) By (MBL1), $\exists x \cdot \exists y \leq \exists(\exists x \cdot \exists y)$. Moreover, since $\exists x \leq \exists y \rightarrow \exists x \cdot \exists y$, then $\exists x = \forall \exists x \leq \forall(\exists y \rightarrow \exists x \cdot \exists y) = \exists y \rightarrow \forall(\exists x \cdot \exists y)$ by Proposition 8(2), 8(14) and (MBL4), which means $\exists x \cdot \exists y \leq \forall(\exists x \cdot \exists y)$. Further, by (MBL5) and (MBL2), $\exists(\exists x \cdot \exists y) \leq \exists \forall(\exists x \cdot \exists y) = \forall(\exists x \cdot \exists y) \leq \exists x \cdot \exists y$.

(2) By (1) and Proposition 8(8), it is immediate.

(3) By (MBL2), $\forall(\forall x \cdot \forall y) \leq \forall x \cdot \forall y$. Moreover, since $\forall x \leq \forall y \rightarrow \forall x \cdot \forall y$, then $\forall x = \forall \forall x \leq \forall(\forall y \rightarrow \forall x \cdot \forall y) = \forall y \rightarrow \forall(\forall x \cdot \forall y)$ by Proposition 8(4), 8(14) and 8(7), so $\forall x \cdot \forall y \leq \forall(\forall x \cdot \forall y)$.

(4) It is immediate by (3) and Proposition 8(8).

(5) On the one hand, $\exists(\exists x \wedge \exists y) \geq \exists x \wedge \exists y$. On the other hand, by Proposition 8(3) and 8(7), we have $\exists(\exists x \wedge \exists y) \leq \exists \exists x = \exists x$ and $\exists(\exists x \wedge \exists y) \leq \exists \exists y = \exists y$, which means $\exists(\exists x \wedge \exists y) \leq \exists x \wedge \exists y$. Therefore, $\exists(\exists x \wedge \exists y) = \exists x \wedge \exists y$.

(6) By (MBL5) and Proposition 8(9), $\exists(\forall x \wedge \forall y) = \exists(\exists \forall x \wedge \exists \forall y) = \forall x \wedge \forall y$. Hence, by Proposition 8(8), (MBL2) and Proposition 8(10), we can get $\forall(\forall x \wedge \forall y) = \forall x \wedge \forall y \leq \forall(x \wedge y) \leq \forall x \wedge \forall y$. Therefore, $\forall(x \wedge y) = \forall x \wedge \forall y$. □ □

Proposition 21. Given an MBSL-algebra $(X, \exists, \forall)$ that satisfies (K). Then the statements as follows are satisfied, for any $x, y \in X$,
(1) $\exists x \cdot \forall y \leq \exists(x \cdot y)$, $\forall x \cdot \exists y \leq \exists(x \cdot y)$,

(2) $\exists(\exists x \cdot y) = \exists x \cdot \exists y = \exists(x \cdot \exists y)$,
(3) $\exists(\forall x \cdot y) = \forall x \cdot \exists y$, $\exists(x \cdot \forall y) = \exists x \cdot \forall y$,
(4) $\forall x \cdot \forall y \leq \forall(x \cdot y) \leq \exists(x \cdot y) \leq \exists x \cdot \exists y$.

Proof. (1) From (MBL2) and Proposition 18(4), $x \cdot \forall y \leq x \cdot y$, then by (MBL1) and Proposition 1, $1 = x \cdot y \to \exists(x \cdot y) \leq x \cdot \forall y \to \forall(x \cdot y)$, so $x \leq \forall y \to \exists(x \cdot y)$. Hence, by Proposition 8(2) and 8(13), $\exists x \leq \exists(\forall y \to \exists(x \cdot y)) = \exists(\forall y \to \forall \exists(x \cdot y)) = \forall y \to \exists(x \cdot y)$, that is, $\exists x \cdot \forall y \leq \exists(x \cdot y)$. Similarly, we can get $\forall x \cdot \exists y \leq \exists(x \cdot y)$.

(2) By (1), we can get $\exists x \cdot \exists y \leq \exists(\exists x \cdot y)$ and $\exists x \cdot \exists y \leq \exists(x \cdot \exists y)$. Further, $\exists x \cdot y \leq \exists x \cdot \exists y$ and $x \cdot \exists y \leq \exists x \cdot \exists y$ yield $\exists(\exists x \cdot y) \leq \exists(\exists x \cdot \exists y) = \exists x \cdot \exists y$ and $\exists(x \cdot \exists y) \leq \exists x \cdot \exists y$ by Proposition 20(1).

(3) By (2), it is immediate.

(4) By (MBL2), Proposition 18(3) and 18(4), $\forall x \cdot \forall y \leq x \cdot y$, then by Propositions 20(3) and 8(14), $\forall(\forall x \cdot \forall y) = \forall x \cdot \forall y \leq \forall(x \cdot y)$. Further, by Proposition 20(3), (MBL1) and (MBL2), $\forall x \cdot \forall y \leq \forall(x \cdot y) \leq \exists(x \cdot y)$. Finally, again using (MBL1), Proposition 18(4) and Proposition 20(1), we have $\exists(x \cdot y) \leq \exists x \cdot \exists y$. □ □

In what follows, we introduce the concepts of relatively complete subalgebras and m-relatively complete subalgebras of self-similar L-algebras. Moreover, a characterization of MBL-algebras and MBSL-algebras is obtained by relatively complete subalgebras and m-relatively complete subalgebras, respectively.

Definition 21. *Given a self-similar L-algebra* $(X, \to, \cdot, 0, 1)$ *and* $A \subseteq X$ *is a subalgebra. We call* A *relatively complete if for any* $x \in X$, $min\{a \in A \mid x \leq a\}$ *and* $max\{a \in A \mid x \geq a\}$ *exist. A relatively complete subalgebra* A *of* X *is called m-relatively complete if for any* $x \in X$ *and* $a \in A$ *with* $x \cdot x \leq a$, *there exists* $b \in A$, *s.t.* $x \leq b$ *and* $b \cdot b \leq a$.

We denote the collection of all relatively complete subalgebras and m-relatively complete subalgebras of X as RC(X) and mRC(X), respectively.

Example 12. If $(X, \exists, \forall)$ is an MBL-algebra, then $X_{\exists\forall} \in RC(X)$. Indeed, if $x \in X$ and $a \in X_{\exists\forall}$ with $x \leq a$, then $x \leq a = \exists a$ iff $\exists x \leq \exists a = a$ by Proposition 8(7), which means $\exists x = min\{a \in X_{\exists\forall} \mid x \leq a\}$. Similarly, $\forall a = a \leq x$ iff $a = \forall a \leq \forall x$, which means $\forall x = max\{a \in X_{\exists\forall} \mid x \geq a\}$. Hence, $X_{\exists\forall} \in RC(X)$.

Example 13. If $(X, \exists, \forall)$ is an MBSL-algebra that satisfies (K), then $X_{\exists\forall} \in mRC(X)$. Indeed, from Example 12, $X_{\exists\forall} \in RC(X)$. Now suppose $x \in X$ and $a \in X_{\exists\forall}$, s.t. $x \cdot x \leq a$, then $a = \exists a \geq \exists(x \cdot x) = \exists x \cdot \exists x \geq x \cdot x$. Taking $b = \exists x$ in Definition 21, then $X_{\exists\forall} \in mRC(X)$.

Theorem 7. Let $(X, \exists, \forall)$ is an MBSL-algebra that satisfies exchange rule, $A \in RC(X)$. If we denote $\exists_A x = min\{a \in A \mid x \leq a\}$ and $\forall_A x = max\{a \in A \mid x \geq a\}$ for any $x \in X$, then $(X, \exists_A, \forall_A)$ is an MBL-algebra.

Proof. (1) Obviously, $\exists_A x = x$ and $\forall_A x = x$ for any $x \in X$, and $\exists_A, \forall_A$ are isotone.

(2) Since $x \leq \exists_A x$ and $\forall_A x \leq x$ for any $x \in X$, then $x \to \exists_A x = 1$ and $\forall_A x \to x = 1$. Hence, (MBL1) and (MBL2) are true.

(3) Let $x, y \in X$, since A is a subalgebra of X and $\exists_A x, \exists_A y \in A$, $\exists_A x \to \exists_A y \in A$. Further, from $\exists_A x \to \exists_A y \leq x \to \exists_A y$, we have $\exists_A x \to \exists_A y \in \{a \in A \mid a \leq x \to \exists_A y\}$. Since $a, \exists_A y \in A$ and $A \subseteq X$, $a \to \exists_A y \in A$. Then for each $a \leq x \to \exists_A y$, exchange rule yields $\exists_A x \leq \exists_A (a \to \exists_A y) = a \to \exists_A y$. Hence, $a \leq \exists_A x \to \exists_A y$, which means $max\{a \in A \mid a \leq x \to \exists_A y\} = \exists_A x \to \exists_A y$. Therefore, $\forall_A (x \to \exists_A y) = \exists_A x \to \exists_A y$, thus (MBL3) is satisfied.

(4) From $\forall_A y \leq y$, we have $\exists_A x \to \forall_A y \leq \exists_A x \to y$, so $\exists_A x \to \forall_A y \in \{a \in A \mid a \leq \exists_A x \to y\}$. If $a \in A$ s.t. $a \leq \exists_A x \to y$, then $a \cdot \exists_A x \leq y$. Since $a \cdot \exists_A x \in A$, $a \cdot \exists_A x = \forall_A (a \cdot \exists_A x) \leq \forall_A y$, hence $a \leq \exists_A x \to \forall_A y$, which means $max\{a \in A \mid a \leq \exists_A x \to y\} = \exists_A x \to \exists_A y$. Therefore, $\forall_A (\exists_A x \to y) = \exists_A x \to \forall_A y$, thus (MBL4) is satisfied.

(5) Since $\forall_A x \in A$, $\exists_A \forall_A x = \forall_A x$, thus (MBL5) is satisfied.

Then we conclude that $(X, \exists_A, \forall_A)$ is an MBL-algebra. □ □

Corollary 4. Given an MBSL-algebra $(X, \exists, \forall)$ that satisfies exchange rule, $A \in mRC(X)$. If we denote $\exists_A x = min\{a \in A \mid x \leq a\}$ and $\forall_A x = max\{a \in A \mid x \geq a\}$ for any $x \in X$, then $(X, \exists_A, \forall_A)$ ia an MBSL-algebra.

Proof. From the proof in Theorem 7, we only need to prove (MBL6) is satisfied. Since $x \cdot x \leq \exists_A x \cdot \exists_A x$, then $\exists_A (x \cdot x) \leq \exists_A x \cdot \exists_A x$ by Proposition 20(1). Moreover, again since $x \cdot x \leq \exists_A x \cdot \exists_A x$ and $\exists_A x \cdot \exists_A x \in A$, then there exists $a \in A$ s.t. $x \leq a$ and $a \cdot a \leq \exists_A x \cdot \exists_A x$. So $x \cdot x \leq a \cdot a \leq \exists_A x \cdot \exists_A x$. Therefore, $\exists_A (x \cdot x) = \exists_A x \cdot \exists_A x$, thus (MBL6) is satisfied. Then we conclude that $(X, \exists_A, \forall_A)$ is an MBSL-algebra. □ □

In what follows, we study the relations between MBSL-algebras and other monadic structures, such as monadic bounded (left) hoops, monadic Wajsberg hoops and monadic MV-algebras.

Definition 22. ([28]) *Given a bounded hoop $(H, \cdot, \to, 0, 1)$. Then we call $(H, \cdot, \to, 0, 1, \exists, \forall)$ of type $(2, 2, 0, 0, 1, 1)$ $((H, \exists, \forall)$ for short) a monadic bounded hoop if for any $x, y \in H$, the statements as follows hold,*
(M1) $x \to \exists x = 1$,

(M2) $\forall x \to x = 1$,
(M3) $\forall(x \to \exists y) = (\exists x \to \exists y)$,
(M4) $\forall(\exists x \to y) = (\exists x \to \forall y)$,
(M5) $\exists \forall x = \forall x$,
(M6) $\forall(x \cdot x) = \forall x \cdot \forall x$.

In Definition 22, if $(H, \cdot, \to, 0, 1)$ is a bounded left hoop, then we call $(H, \exists, \forall)$ a *monadic bounded left hoop*.

Theorem 8. *([15]) Every self-similar L-algebra is a left hoop.*

Remark 4. If the self-similar L-algebra X satisfies exchange rule, that is, $x \to (y \to z) = y \to (x \to z)$, then X becomes a hoop. Indeed, if X satisfies exchange rule, then by Proposition 2(2), we have $x \cdot y \to z = y \cdot x \to z$, which means $x \cdot y = y \cdot x$. Hence, the binary operation $\cdot$ on X is commutative, so X becomes a hoop.

Theorem 9. Given an MBSL-algebra $(X, \exists, \forall)$ that satisfies (K). Then $(X, \exists, \forall)$ is a monadic bounded left hoop if for any $x \in X$, $\forall(x \cdot x) \leq \forall x \cdot \forall x$.

Proof. From Theorem 8, X is a bounded left hoop. Moreover, axioms (M1) to (M5) are consistent with (MBL1) to (MBL5), respectively. So we only need to verify (M6). From Proposition 18(3) and 18(4), we have $\forall x \cdot \forall x \leq x \cdot x$ and hence $\forall x \cdot \forall x = \forall(\forall x \cdot \forall x) \leq \forall(x \cdot x)$ by Proposition 20(3). Therefore, (M6) is true. Then we conclude $(X, \exists, \forall)$ is a monadic bounded left hoop. □ □

Corollary 5. Given an MBSL-algebra $(X, \exists, \forall)$ that satisfies exchange rule. Then $(X, \exists, \forall)$ is a monadic bounded hoop if for any $x \in X$, $\forall(x \cdot x) \leq \forall x \cdot \forall x$.

Proof. From Remark 3 and Theorem 9, it is immediate. □

Definition 23. *([4]) Given a Wajsberg hoop $(H; \cdot, \to, 1)$ with a unary operator $\forall$ (an universal quantifier). We call $(H, \forall)$ monadic Wajsberg hoop if the statements as follows hold, for any $x, y \in H$,*
(MH1) $\forall 1 = 1$,
(MH2) $\forall x \to x = 1$,
(MH3) $\forall((x \to \forall y) \to \forall y) = (\forall x \to \forall y) \to \forall y$,
(MH4) $\forall(x \to y) \to (\forall x \to \forall y) = 1$,
(MH5) $\forall(\forall x \to \forall y) = \forall x \to \forall y$,
(MH6) $\forall(x \cdot x) = \forall x \cdot \forall x$,
(MH7) $\forall((x \to \forall y) \to x) = (\forall x \to \forall y) \to \forall x$,
(MH8) $\forall(x \wedge y) = \forall x \wedge \forall y$,
(MH9) $\forall(\forall x \cdot \forall y) = \forall x \cdot \forall y$.

Remark 5. From the proof of statements (18) and (19) in Proposition 8, we can get that (MH3) ⇒ (MH7). So we can say that a Wajsberg hoop $(H; \cdot, \rightarrow, \forall, 1)$ is called a monadic Wajsberg hoop if for any $x, y \in H$, axioms (MH1)-(MH6), (MH8) and (MH9) are hold.

Proposition 22. Every self-similar L-algebra that satisfies (C) and exchange rule is a Wajsberg hoop.

Proof. From Remark 3 and Definition 6, it is immediate. □ □

Theorem 10. Given an MBSL-algebra $(X, \exists, \forall)$ that satisfies (C) and exchange rule. Then $(X, \exists, \forall)$ is a monadic Wajsberg hoop if for any $x \in X$, $\forall (x \cdot x) \leq \forall x \cdot \forall x$.

Proof. From Proposition 22, X is a Wajsberg hoop. Moreover, axioms (MH1)-(MH5), (MH8)-(MH9) are consistent with Proposition 8(1), (MBL2), Proposition 8(18), 8(16), 8(11), Proposition 20(6) and 20(3), respectively. And the proof of (MH6) is similar to Theorem 9, so we omit it. □ □

Definition 24. ([8, 9]) *Given an MV-algebra M with a unary operator $\forall$ (an universal quantifier). We call $(M, \forall)$ monadic MV-algebra if for any $x, y \in M$, the statements as follows are satisfied,*
(A1) $x \geq \forall x$,
(A2) $\forall (x \wedge y) = \forall x \wedge \forall y$,
(A3) $\forall (\forall x)' = (\forall x)'$,
(A4) $\forall (\forall x \cdot \forall y) = \forall x \cdot \forall y$,
(A5) $\forall (x \cdot x) = \forall x \cdot \forall x$,
(A6) $\forall (x \oplus x) = \forall x \oplus \forall x$.

From Propositions 4 and 22, we have the following corollary.

Corollary 6. Every self-similar L-algebra that satisfies (C) and exchange rule is an MV-algebra.

Given a bounded self-similar L-algebra $(X, \cdot, \rightarrow, 0, 1)$, for any $x, y \in X$, we define $x \cdot y = (x \rightarrow y')'$. Then $(X, \cdot, ', 1)$ is an MV-algebra, where $x' = x \rightarrow 0$ ([31]). For any bounded self-similar L-algebra X and $x \in L$, besides the operation $x' = x \rightarrow 0$, we also define $x \oplus y = (x' \cdot y')'$.

Theorem 11. Given a strong MBSL-algebra $(X, \exists, \forall)$ that satisfies (C) and exchange rule. Then $(X, \exists, \forall)$ is a monadic MV-algebra if for any $x \in X$, $\forall (x \cdot x) \leq \forall x \cdot \forall x$.

Proof. From Corollary 6, X is an MV-algebra. Moreover, axioms (A1)-(A4) are consistent with (MBL2), Propositions 20(6), 9(2) and 20(3), respectively. The proof of (A5) is similar to Theorem 9. So we only need to prove (A6). From Proposition 9(3), (MBL6) and Proposition 9(4), we can get $\forall (x \oplus x) = \forall ((x' \cdot x')') = (\exists (x' \cdot x'))' = (\exists x' \cdot \exists x')' = (\exists x')' \oplus (\exists x')' = \forall x \oplus \forall x$. Therefore, $(X, \exists, \forall)$ is a monadic MV-algebra. □

□

7 Conclusion

As we all known, many monadic algebras have been researched in the literature. Inspired and based on this, we spread the notion of monadic operators to a more general algebraic structure, namely L-algebras. As we said in introduction, L-algebras are also logical algebras. So in the next work, we will discuss the construction, properties and completeness of monadic formal deductive system corresponding to monadic L-algebras class, so as to expand the research of L-algebras and corresponding logic system.

Acknowledgements 1. This study was funded by grants of National Natural Science Foundation of China (grant number 11971384) and Natural Science Foundation of Shaanxi Province (grant number 2021JQ-894).

Conflict of interest

The authors declare that they have no conflict of interest.

References

[1] Blyth, T.S.: Lattices and ordered algebraic structures. Springer-Verlag London Limited.(2005)

[2] Bosbach, B.: Rechtskomplementäre halbgruppen: Axiome, polynome, kongruenzen. Mathematische Zeitschrift 124(4), 273-288(1972)

[3] Chang, C.C.: Algebraic analysis of many valued logics. Transactions of the American Mathematical Society 88(2), 467-490(1958)

[4] Cimadamore, C.R., Varela, J.: Monadic Wajsberg hoops. Revista de la Union Matematica Argentina 57(2), 63-83(2016)

[5] Ciungu, L.C.: Results in L-algebras. Algebra universalis 82(1)(2020)

[6] Ciungu, L.C.: Monadic classes of quantum B-algebras. Soft Computing 25, 1-14(2021)

[7] Ciungu, L.C.: Quantifiers on L-algebras. Math. Slovaca 72, 1403-1428(2022)

[8] Di Nola, A., Grigolia, R.: On monadic MV-algebras. Annals of pure and applied logic 128, 125-139(2004)

[9] Figallo, O.A.: A topological duality for monadic MV-algebras. Soft Computing 21, 7119-7123(2017)

[10] Halmos, P.R.: Algebraic logic. New York: Chelsea Publishing Company(1962)

[11] Herman, L., Marsden, E.L., Piziak, R.: Implication connectives in orthomodular lattices. Notre Dame Journal of Formal Logic 16, 305-328(1975)

[12] Iorgulescu, A.: Monadic involutive pseudo-BCK algebras. Acta Universitatis Apulensis 15, 159-178(2008)

[13] Nemeti, I.: Algebraization of quantifier logics, an introductory overview. Studia Logica 50, 485-496(1991)

[14] Pigozzi, D., Salibra, A.: Polyadic algebras over nonclassical logics. Algebraic Methods in Logic and in Computer Science 28, 51-56(1993)

[15] Rump, W.: L-algebras, self-similarity, and l-groups. Journal of Algebra 320, 2328-2348(2008)

[16] Rump, W.: Semidirect products in algebraic logic and solution of the quantum Yang-Baxter equation. Journal of Algebra and Its Applications 7(4), 471-490(2008)

[17] Rump, W.: A decomposition theorem for square-free unitary solutions of the quantum Yang-Baxter equation. Advances in Mathematics 193, 40-55(2005)

[18] Rump, W.: Braces, radical rings, and the quantum Yang-Baxter equation. Journal of Algebra 307, 153-170(2007)

[19] Rump, W.: Quantum B-algebra. Central European Journal of Mathematics 1811-1899(2013)

[20] Rump, W., Yang, Y.C.: Non-commutative logic algebras and algebraic quantales. Annals of Pure and Applied Logic 165, 759-785(2014)

[21] Rump, W.: Symmetric quantum sets and L-algebras. International Mathematics Research Notices 00, 1-41(2020)

[22] Rump, W.: Right l-groups, geometric Garside groups, and solutions of the quantum Yang-Baxter equation. Journal of Algebra 439, 470-510(2015)

[23] Rump, W.: The structure group of an L-algebra is torsion-free. Group Theory 2, 309-324(2017)

[24] Rump, W.: Von neumann algebras, L-algebras, Baer*-monoids, and Garside groups. Forum Mathematicum 3, 973-995(2018)

[25] Rump, W.: The structure group of a generalized orthomodular lattice. Studia Logica 106, 85-100(2018)

[26] Rump, W.: L-effect Algebras. Studia Logica 108, 725-75(2019)

[27] Traczky, T.: On the structure of BCK-algebras with $zx \cdot yx = zy \cdot xy$. Osaka Journal of Mathematics 33(2), 319-324(1988)

[28] Wang, J.T., Xin, X.L., He, P.F.: Monadic bounded hoops. Soft Computing 22, 1749-1762(2018)

[29] Wang, J.T., He, P.F., She, Y.H.,: Monadic NM-algebras, Logic Journal of the IGPL 27, 812-835(2019)

[30] Wang, J.T., He, P.F., Yang J., Wang M., He X.L., Monadic NM-algebras: an algebraic approach to monadic predicate nilpotent minimum logic, Journal of Logic and Computation 32, 741-766(2022)

[31] Wu, Y., Wang, J., Yang, Y.: Lattice-ordered effect algebras and L-algebras. Fuzzy Sets and Systems 369, 103-113(2019)

[32] Xin, X.L., Fu, Y.L., Lai, Y., Wang, J.T.: Monadic pseudo BCI-algebras and corresponding logics. Soft Computing 23, 1499-1510(2019)

Ideals on pseudo equality algebras

Zhaoping Lu [a]
a. School of Science, Xi'an Polytechnic University, Xi'an 710048, P.R. China
2353394464@qq.com

Xiaolong Xin [a,b,*]
a. School of Science, Xi'an Polytechnic University, Xi'an 710048, P.R. China
b. School of Mathematics, Northwest University, Xi'an 710127, P.R. China
xlxin@nwu.edu.cn

Abstract

We mainly focus on studying ideals on pseudo equality algebras. Firstly, we introduce the concept of ideals on pseudo equality algebras and provide some examples. We discuss their related properties and give the equivalent characterization of ideals. Next, we explore the relationships between ideals and filters on pseudo equality algebras under certain conditions and study the generation formula of ideals on involutive pseudo equality algebras. Then we induce congruence relations by ideals and construct the quotient structures. In addition, we introduce the concept of prime ideals on pseudo equality algebras and research their related properties. We provide the equivalent characterizations of prime ideals. We prove that if a pseudo equality algebra is a chain, then each proper ideal is a prime ideal, but the converse is not true. Finally, we introduce the concept of maximal ideals on pseudo equality algebras and discuss the relationships between maximal ideals and prime ideals.

Keywords: Pseudo equality algebra; ideal; congruence; prime ideal

1 Introduction

In recent years, the study of fuzzy logic has become a hot topic in information science research, and fuzzy logic research is inseparable from logic algebra. Various logic algebras have been introduced and studied as non-classical logic semantic systems, such as residuated lattice [18], BL-algebras [11], MV-algebras [2] and Weak Pseudo

---

*Sponsored by a grant of National Natural Science Foundation of China (11971384)

EMV-algebras [8]. In order to develop the truth algebraic structure of fuzzy type theory, V. Novák and B.De Baets [15] proposed a new logical algebra named EQ-algebras in 2009. EQ-algebras have three fundamental operations, meet operation $\wedge$, product operation $\otimes$ and fuzzy equality operation $\sim$, implication $\rightarrow$ can be induced by fuzzy equality operation $\sim$. As soon as EQ-algebra was founded, it attracted the attention and research of many scholars at home and abroad, and obtained many important conclusions [9, 1, 20]. However, it was founded that the product operation in EQ-algebra was still an EQ-algebra after being replaced by another smaller binary operation. Therefore, S. Jenei [12] proposed a new algebraic structure named equality algebras in 2012. Compared with EQ-algebra, equality algebra has no product operation. As a generalization of equality algebras, Jenei [13] put forward the concept of pseudo equality algebras and proved that the pseudo equality algebras and the pseudo BCK-meet-semilattices are equivalent in 2013. In 2014, Ciungu [4] discovered a gap in the above equivalence proof of [13], gave a counterexample and the correct version of the theorem. In addition, Dvurečenskij and Zahiri [7] proved that each pseudo equality algebra in [13] is equality algebra and introduced a new version of pseudo equality algebras in 2016. From a logical point of view, the study of equality algebras and pseudo equality algebras is meaningful and can also enrich the general algebraic system, pseudo equality algebras are not the same as other algebraic systems, and as a generalization of equality algebras, the introduction of $\rightsquigarrow$ will further restrict the pseudo equality algebras.

Ideals are an important way to study logical algebras. Many logical algebras have proposed the concept of ideals, for example pseudo-MV algebras [10], pseudo-hoop algebras [19] and equality algebras [16]. Georgescu and Iorgulescu [10] proposed the concept of ideals on pseudo-MV algebras and introduced prime ideals and normal ideals, which have been proven to be effective in studying the structural properties of pseudo-MV algebras. In addition, Wenjuan Chen [5] studied ideals and congruence relations on quasi-pseudo-MV algebras. Fei Xie and Hongxing Liu [19] studied ideals on pseudo-hoop algebras, induced the congruence relations by ideals, and constructed quotient structures. Jie Qiong Shi and Xiao Long Xin [17] introduced the concept of ideals on EQ-algebras, studied relevant properties and equivalent characterizations. They [17] also discussed the properties and relations of implicative ideals, primary ideals, prime ideals and maximal ideals. Akbar Paad [16] introduced the concept of ideals on bounded equality algebras and studied relevant properties. Furthermore, Paad [16] introduced prime ideals and Boolean ideals on bounded equality algebras. We find that the concept of ideals has not be introduced on the pseudo equality algebras, and there are few examples of pseudo equality algebras, which may bring difficulties in study the algebraic structure of logic systems. The pseudo equality algebra is a simplification of the equality algebra, and we would like to investigate the differences between the two algebraic structures by studying the differences in ideals.

This article is structured as follows: In Section 2, we review some concepts and relevant properties of pseudo equality algebras. In Section 3, we introduce the concept of ideals, give several examples of pseudo equality algebras, and provide the equivalent characterization of ideals and generating formula. We also discuss the relationship between ideals and filters of pseudo equality algebras. In addition, we induce the congruence relations by ideals. In Section 4, we introduce prime ideals and maximal ideals, provide equivalent characterizations of prime ideals, and discuss relevant properties.

2 Preliminaries

Below are the definitions and basic results of pseudo equality algebras.

Definition 2.1. *[14] An algebra $M = (M, \wedge, \sim, \backsim, 1)$ of type $(2,2,2,0)$ is said to be a pseudo equality algebra (or a JK-algebra) if it fulfills the following conditions, for each $s, t, w \in M$,*
$(M1)$ $(M, \wedge, 1)$ *is a meet-semilattice with top element 1,*
$(M2)$ $s \sim s = s \backsim s = 1$,
$(M3)$ $s \sim 1 = 1 \backsim s = s$,
$(M4)$ $s \leq t \leq w$ *implies* $s \sim w \leq t \sim w, s \sim w \leq s \sim t, w \backsim s \leq w \backsim t$ *and* $w \backsim s \leq t \backsim s$,
$(M5)$ $s \sim t \leq (s \wedge w) \sim (t \wedge w)$ *and* $s \backsim t \leq (s \wedge w) \backsim (t \wedge w)$,
$(M6)$ $s \sim t \leq (w \sim s) \backsim (w \sim t)$ *and* $s \backsim t \leq (s \backsim w) \sim (t \backsim w)$,
$(M7)$ $s \sim t \leq (s \sim w) \sim (t \sim w)$ *and* $s \backsim t \leq (w \backsim s) \backsim (w \backsim t)$.

We refer to $\sim$ and $\backsim$ as pseudo equality operations and $\wedge$ as the meet operation. We define $s \leq t$, for all $s, t \in M$, by $s \wedge t = s$. Two unary operations are defined, $s^- = s \to 0$, $s^\sim = s \rightsquigarrow 0$, for all $s \in M$. Also, we define two other operations, called implications,
$$s \to t = (s \wedge t) \sim s,$$
$$s \rightsquigarrow t = s \backsim (s \wedge t).$$

Definition 2.2. *[14, 3] Let M be a pseudo equality algebra. For all $s, t \in M$, we call M is,*
(1) *bounded if there is an element $0 \in M$, $0 \leq s$,*
(2) *involutive if $s^{\sim -} = s, s^{-\sim} = s$,*

(3) good if $s^{-\sim} = s^{\sim-}$,
(4) symmetric if $s \backsim t = t \sim s$.

Proposition 2.3. *[14] Let M be a pseudo equality algebra. For all $s, t, w \in M$, the following properties hold,*
(M8) $s \backsim t \leq s \rightsquigarrow t$ and $t \sim s \leq s \rightarrow t$,
(M9) $s \leq ((w \sim s) \backsim w) \wedge (w \sim (s \backsim w))$, $s \leq ((s \backsim w) \sim (1 \backsim w)) \wedge ((w \backsim 1) \sim (w \backsim s))$, $s \backsim 1 \leq ((s \backsim w) \sim w) \wedge ((w \backsim s) \sim (w \backsim 1))$, $1 \sim s \leq (w \backsim ((w \backsim s)) \wedge ((1 \sim w) \sim (s \sim w))$,
(M10) $s \backsim t = 1$ or $t \sim s = 1$ imply $s \leq t$,
(M11) $s \sim t = 1$ implies $w \sim s \leq w \sim t$ and $s \backsim t = 1$ implies $t \backsim w \leq s \backsim w$,
(M12) $s \leq t$ iff $s \rightarrow t = 1$ iff $s \rightsquigarrow t = 1$,
(M13) $s \rightsquigarrow 1 = s \rightsquigarrow s = s \rightarrow s = s \rightarrow 1 = 1, 1 \rightsquigarrow s = s$ and $1 \rightarrow s = s$,
(M14) $s \leq (t \rightarrow s) \wedge (t \rightsquigarrow s)$,
(M15) $s \leq ((s \rightarrow t) \rightsquigarrow t) \wedge ((s \rightsquigarrow t) \rightarrow t)$,
(M16) $s \rightarrow t \leq (t \rightarrow w) \rightsquigarrow (s \rightarrow w)$ and $s \rightsquigarrow t \leq (t \rightsquigarrow w) \rightarrow (s \rightsquigarrow w)$,
(M17) $s \leq t \rightarrow w$ iff $t \leq s \rightsquigarrow w$,
(M18) $s \rightarrow (t \rightsquigarrow w) = t \rightsquigarrow (s \rightarrow w)$,
(M19) $t \rightarrow s \leq (t \wedge w) \rightarrow (s \wedge w)$ and $t \rightsquigarrow s \leq (t \wedge w) \rightsquigarrow (s \wedge w)$,
(M20) $s \rightarrow t = s \rightarrow (s \wedge t)$ and $s \rightsquigarrow t = s \rightsquigarrow (s \wedge t)$,
(M21) $1 \sim s = s \backsim 1$,
(M22) if $s \leq t$, then $s \leq (t \backsim s) \wedge (s \sim t)$,
(M23) $s \backsim t \leq 1 \sim (t \backsim s)$ and $s \sim t \leq 1 \sim (t \sim s)$,
(M24) $t \leq w$ implies $s \rightarrow t \leq s \rightarrow w$ and $s \rightsquigarrow t \leq s \rightsquigarrow w$,
(M25) $t \leq w$ implies $w \rightarrow s \leq t \rightarrow s$ and $w \rightsquigarrow s \leq t \rightsquigarrow s$.

Definition 2.4. *[14] Let M be a pseudo equality algebra and $J \subseteq M$. J is called a filter, if for all $u, v \in M$, it fulfills the following,*
(1) $1 \in J$,
(2) $s \in J, s \leq t$ imply $t \in J$,
(3) $s, s \backsim t \in J$ imply $t \in J$.
Condition (3) is equivalent to the condition,
(3') $s, t \sim s \in J$ imply $t \in J$.

Define the set of all filters of M by $F(M)$.

Proposition 2.5. *[14] Let M be a pseudo equality algebra. The following conditions are equivalent,*
(1) $J \in F(M)$,
(2) for all $s, t \in M$, $1 \in J$ and $s, s \rightarrow t \in J$ imply $t \in J$,

(3) for all $s,t \in M$, $1 \in J$ and $s, s \rightsquigarrow t \in J$ imply $t \in J$.

Definition 2.6. *[14] Let M be a pseudo equality algebra and $M \neq K \in F(M)$. If for all $s,t \in M$, $s \to t \in K$ or $t \to s \in K$, $s \rightsquigarrow t \in K$ or $t \rightsquigarrow s \in K$, then we call K a prime filter.*

Proposition 2.7. *[14] Let M be a pseudo equality algebra. For all $s,t,w \in M$, the following properties hold,*
(1) $s \to t \leq (w \to s) \to (w \to t)$,
(2) $s \rightsquigarrow t \leq (w \rightsquigarrow s) \rightsquigarrow (w \rightsquigarrow t)$.

Definition 2.8. *[14] A lattice pseudo equality algebra is a pseudo equality algebra in which $(X, \leq)$ is a lattice, as well.*

Proposition 2.9. *[14] Let M be a lattice pseudo equality algebra. For all $s,t,w \in M$, the following properties hold,*
(1) *For all indexed families $\{s_i\}_{i \in I}$ in M, we have $(\vee_{i \in I} s_i) \to t = \wedge_{i \in I}(s_i \to t)$ and $(\vee_{i \in I} s_i) \rightsquigarrow t = \wedge_{i \in I}(s_i \rightsquigarrow t)$, provided that the infimum and supremum of $\{s_i\}_{i \in I}$ exist in M,*
(2) $(s \vee t) \to w = (s \to w) \wedge (t \to w)$ and $(s \vee t) \rightsquigarrow w = (s \rightsquigarrow w) \wedge (t \rightsquigarrow w)$,
(3) $s \to t = (s \vee t) \to t$ and $s \rightsquigarrow t = (s \vee t) \rightsquigarrow t$.

Proposition 2.10. *Let M be a lattice pseudo equality algebra. For all $s,t \in M$, the following properties hold,*
(1) $s \leq s^{-\sim}$, $s \leq s^{\sim -}$,
(2) $(s \vee t)^- = s^- \wedge t^-$, $(s \vee t)^\sim = s^\sim \wedge t^\sim$,
(3) $s^- = s^{-\sim -}$, $s^\sim = s^{\sim -\sim}$,
(4) $(s^- \to t^-)^{\sim -} = s^- \to t^-$, $(s^\sim \rightsquigarrow t^\sim)^{-\sim} = s^\sim \rightsquigarrow t^\sim$.

Proof. (1) By (M15), $s \leq ((s \to 0) \rightsquigarrow 0) \wedge ((s \rightsquigarrow 0) \to 0) = s^{-\sim} \wedge s^{\sim -}$, then we $s \leq s^{-\sim}$, $s \leq s^{\sim -}$.
(2) By Proposition 2.9(2), $(s \vee t) \to w = (s \to w) \wedge (t \to w)$, $(s \vee t) \rightsquigarrow w = (s \rightsquigarrow w) \wedge (t \rightsquigarrow w)$. Thus $(s \vee t)^- = (s \to 0) \wedge (t \to 0) = s^- \wedge t^-$, $(s \vee t)^\sim = (s \rightsquigarrow 0) \wedge (t \rightsquigarrow 0) = s^\sim \wedge t^\sim$.
(3) By (M25) and (M18), $s^{-\sim -} \leq s^-$, $s^- \rightsquigarrow s^{-\sim -} = s^{-\sim} \to s^{-\sim} = 1$. Then $s^- \leq s^{-\sim -}$, $s^- = s^{-\sim -}$. Similarly, we get that $s^\sim = s^{\sim -\sim}$.
(4) By (1), $s^- \to t^- \leq (s^- \to t^-)^{\sim -}$. By (M18), $(s^- \to t^-)^{\sim -} \rightsquigarrow (s^- \to t^-) = s^- \to ((s^- \to t^-)^{\sim -} \rightsquigarrow t^-) = s^- \to (t^- \to (s^- \to t^-)^{\sim - \sim}) = s^- \to (t^- \to (s^- \to t^-)^\sim) = s^- \to ((s^- \to t^-) \rightsquigarrow t^-) = (s^- \to t^-) \rightsquigarrow (s^- \to t^-) = 1$. Thus $(s^- \to t^-)^{\sim -} \leq s^- \to t^-$. Therefore, $(s^- \to t^-)^{\sim -} = s^- \to t^-$. Also, we can prove $(s^\sim \rightsquigarrow t^\sim)^{-\sim} = s^\sim \rightsquigarrow t^\sim$. □

Definition 2.11. *[14] Let M be a pseudo equality algebra. For all $s, t \in M$, if 1 is a unique upper bound of the set $\{s \to t, t \to s\}$ and $\{s \rightsquigarrow t, t \rightsquigarrow s\}$, i.e., $(s \to t) \vee (t \to s) = 1 = (s \rightsquigarrow t) \vee (t \rightsquigarrow s)$, then we call M is prelinear.*

Proposition 2.12. *[14] Let M be a prelinear pseudo equality algebra. For all $s, t \in M$, the following properties hold,*
(1) $(s \wedge t) \to w = (s \to w) \vee (t \to w)$ and $(s \wedge t) \rightsquigarrow w = (s \rightsquigarrow w) \vee (t \rightsquigarrow w)$,
(2) $s \to (t \wedge w) = (s \to t) \wedge (s \to w)$ and $s \rightsquigarrow (t \wedge w) = (s \rightsquigarrow t) \wedge (s \rightsquigarrow w)$.

Definition 2.13. *[7] Let M be a pseudo equality algebra and $g \in M$. The g is called invariant, if it fulfills the condition,*

$$g \leq t \text{ implies } g \backsim t = 1 = t \sim g \text{ for all } t \in M.$$

3 Ideals on pseudo equality algebras

In this section, we will introduce the definition of ideals and their generating formula, as well as discuss the relationships between ideals and filters. Additionally, we will derive congruence relations by ideals.

Definition 3.1. *Let M be a bounded pseudo equality algebra, $\varnothing \neq T \subseteq M$. The T is called a left ideal, if it fulfills the following conditions,*
(MT1) for every $s, t \in M$, $s \leq t$ and $t \in T$ imply $s \in T$,
(MT2) for every $s, t \in T$, $s^\sim \to t \in T$.

T is called a right ideal, if it fulfills the following conditions,
(MT1) for every $s, t \in M$, $s \leq t$ and $t \in T$ imply $s \in T$,
(MT3) for every $s, t \in T$, $s^- \rightsquigarrow t \in T$.

If T is the left ideal of M and also the right ideal of M, we call T an ideal of M.

Suppose that M is a pseudo equality algebra. Denote the set of all ideals of M by $I(M)$. If $M \neq T \in I(M)$, then T is said to be a proper ideal.

Example 3.2. *[7] Let $M = \{0, i, j, 1\}$ be a chain with $0 < i < j < 1$. The operations $\backsim$ and $\sim$ on M given as follows,*

Table 1: Cayley table for the binary operation "$\sim$"

$\sim$	0	i	j	1
0	1	j	j	0
i	1	1	j	i
j	1	1	1	j
1	1	1	1	1

Table 2: Cayley table for the binary operation "⌣"

⌣	0	i	j	1
0	1	1	1	1
i	j	1	1	1
j	0	i	1	1
1	0	i	j	1

After calculations, we can observe that M is a JK-algebra. $T_1 = \{0\}$, $T_2 = \{0, i\}$ and M are the ideals.

Let $\{T_\lambda : \lambda \in \Lambda\}$ be a family of ideals, then $\cap_{\lambda \in \Lambda} T_\lambda$ is an ideal, however $\cup_{\lambda \in \Lambda} T_\lambda$ is not necessarily an ideal in general.

Example 3.3. Let $M = \{0, i, j, 1\}$ in which the Hasse diagram and the operations $⌣$ and $\sim$ on M given as follows,

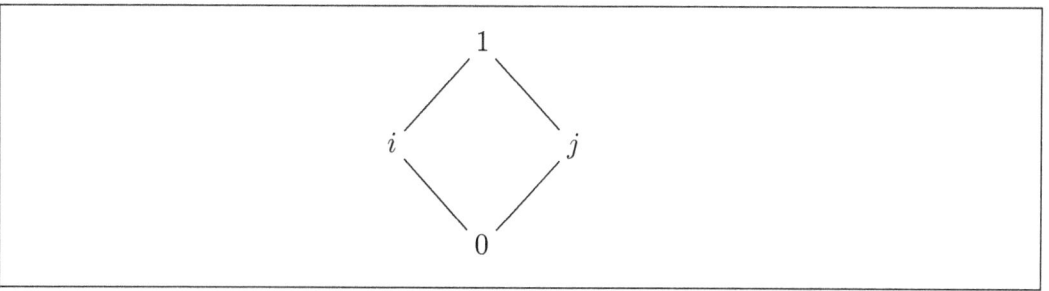

Figure 1: Hasse Diagram of M

Table 3: Cayley table for the binary operation "$\sim$"

$\sim$	0	i	j	1
0	1	j	i	0
i	j	1	j	i
j	i	j	1	j
1	0	i	j	1

⌣	0	i	j	1
0	1	j	1	j
i	j	1	j	i
j	i	i	1	j
1	0	i	j	1

Table 4: Cayley table for the binary operation "⌣"

After calculations, we can observe that M is a JK-algebra. $T_1 = \{0\}$, $T_2 = \{0, i\}$, $T_3 = \{0, j\}$ and $T_4 = M$ are the ideals. But $T_1 \cup T_2 = \{0, i, j\}$ is not an ideal, since $i^\sim \to j = j \to j = 1 \notin T_1 \cup T_2$ and $j^- \leadsto i = i \leadsto i = 1 \notin T_1 \cup T_2$.

Example 3.4. Let $M = \{0, i, j, k, 1\}$ in which the Hasse diagram and the operations $\smile$ and $\sim$ on M given as follows,

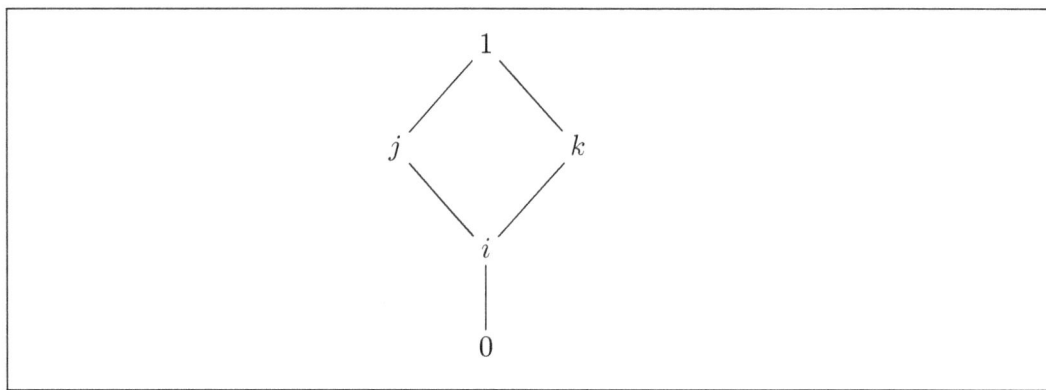

Figure 2: Hasse Diagram of M

Table 5: Cayley table for the binary operation "$\sim$"

$\sim$	0	i	j	k	1
0	1	k	0	j	0
i	1	1	i	j	i
j	1	1	1	j	j
k	1	1	k	1	k
1	1	1	1	1	1

$\smile$	0	i	j	k	1
0	1	1	1	1	1
i	j	1	1	1	1
j	i	k	1	k	1
k	i	i	j	1	1
1	0	i	j	k	1

Table 6: Cayley table for the binary operation "$\rightsquigarrow$"

After calculations, we can observe that M is a JK-algebra. $T_1 = \{0\}$, $T_2 = \{0, i\}$ are the ideals. $T_3 = \{0, i, j\}$ is not an ideal, since $j^\sim \to j = j \to j = 1 \notin T_3$ and $j^- \rightsquigarrow j = 0 \rightsquigarrow j = 1 \notin T_3$. $T_4 = \{0, i, k\}$ is not an ideal, since $k^\sim \to k = a \to k = 1 \notin T_4$ and $k^- \rightsquigarrow k = j \rightsquigarrow k = k \in T_4$.

Theorem 3.5. *Let M be a bounded good pseudo equality algebra and $T \subseteq M$. Then $T \in I(M)$ iff it fulfills the following conditions,*
(MT4) $0 \in T$,
(MT5) for every $s, t \in M$, $s \in T$ and $(s^- \rightsquigarrow t^-)^\sim \in T$ imply $t \in T$.
$T \in I(M)$ iff it fulfills (MT4) and the condition,
(MT6) for every $s, t \in M$, $s \in T$ and $(s^\sim \to t^\sim)^- \in T$ imply $t \in T$.

Proof. ($\Rightarrow$) Suppose that $T \in I(M)$. Since $\emptyset \neq T \subseteq M$, then there exists $s \in M$, such that $s \in T$. Since M is bounded, then we have $0 \leq s$. By (MT1), $0 \in T$. Assume that $s, (s^- \rightsquigarrow t^-)^\sim \in T$. By (MT2), $(s^\sim \to (s^- \rightsquigarrow t^-)^\sim) \in T$. By (M18), $((s^- \rightsquigarrow t^-) \rightsquigarrow s^{\sim -}) \in T$.

$$\begin{aligned} t \to ((s^- \rightsquigarrow t^-) \rightsquigarrow s^{\sim -}) &= (s^- \rightsquigarrow t^-) \rightsquigarrow (t \to s^{\sim -}) \\ &= (s^- \rightsquigarrow t^-) \rightsquigarrow (t \to s^{-\sim}) \\ &= (s^- \rightsquigarrow t^-) \rightsquigarrow (s^- \rightsquigarrow t^-) \\ &= 1. \end{aligned}$$

By (M12), $t \leq (s^- \rightsquigarrow t^-) \rightsquigarrow s^{\sim -}$. By (MT1), we can get that $t \in T$.
($\Leftarrow$) Let $s \leq t$ and $t \in T$, for $s, t \in M$. Then $t^- \leq s^-$, thus $(t^- \rightsquigarrow s^-)^\sim = 0 \in T$. Since $t \in T$ and (MT5), we have $s \in T$. If $s^{\sim -} \in T$ and $T \in I(M)$, then $s \in T$ by Proposition 2.10. Let $s \in T$, by (M12),(M13) and (M15), $(s^\sim \to s^{\sim - \sim})^- = 1^- = 0 \in T$, we conclude that $s^{\sim -} \in T$. Let $s, t \in T$, by (M15) and (M16),

$$(t^- \rightsquigarrow (s^\sim \to t)^-)^- \leq ((s^\sim \to t) \rightsquigarrow t)^- \leq s^{\sim -}.$$

Since $s \in T$, we have $s^{\sim -} \in T$ and so $(t^- \rightsquigarrow (s^\sim \to t)^-)^- \in T$. By (MT2), $s^\sim \to t \in T$. Similarly, by (MT3), we get that (MT6). $\square$

Let $(M, \wedge, \sim, \rightsquigarrow, 1)$ be a bounded pseudo equality algebra and $Q \subseteq M$. We denote,

$$D(Q) = \{s \in M | s^- \in Q\}, \quad E(Q) = \{s \in M | s^\sim \in Q\}$$

Proposition 3.6. *Let M be a bounded pseudo equality algebra. Then the following hold,*
(1) *if M is good and $T \in I(M)$, then $D(T)$ and $E(T)$ are the filters of M,*
(2) *if M is involutive and $J \in F(M)$, then $D(J)$ and $E(J)$ are the ideals of M.*

Proof. (1) By $(MT4)$, $0 \in T$. Since $1 = 0^-$, we have $1 \in D(T)$. Let $s, s \to t \in D(T)$, for $s, t \in M$. Then $(s \to t)^- \in T$, $s^- \in T$. By $(M16)$, $(s \to t) \leq (t \to 0) \rightsquigarrow (s \to 0)$ and we also have $(t^- \rightsquigarrow s^-)^- \leq (s \to t)^- \in T$. Hence, $(t^- \rightsquigarrow s^-)^- \in T$. By Proposition 2.10, we have $(t^- \rightsquigarrow s^{-\sim -})^- \in T$. By $(M18)$, $(s^{-\sim} \to t^{-\sim})^- \in T$. Since $s^- \in T$ and $T \in I(M)$, we get that $t^- \in T$ by $(MT5)$. Thus $t \in D(T)$. We can conclude that $D(T) \in F(M)$. By the similar way, let $s, s \rightsquigarrow t \in M(T)$, for $s, t \in M$, we have $t^\sim \in T$. Hence, $t \in M(T)$. We can conclude that $E(T) \in F(M)$.
(2) Since $1 \in J$, we have $0 = 1^- \in D(J)$. Let $s, (s^- \rightsquigarrow t^-)^\sim \in D(J)$, for $s, t \in M$. Then $s^- \in J$ and $(s^- \rightsquigarrow t^-)^{\sim -} \in J$. Since M is involutive, we have $s^- \rightsquigarrow t^- = (s^- \rightsquigarrow t^-)^{\sim -}$. Thus $s^- \rightsquigarrow t^- \in J$. Since $J \in F(M)$ and $s^- \in J$, we get that $t^- \in J$. Hence, $t \in D(J)$. We can conclude that $D(J) \in I(M)$. By the similar way, let $s, (s^\sim \to t^\sim)^- \in E(J)$, for $s, t \in M$, we have $t^\sim \in J$. Hence, $t \in E(J)$. We can conclude that, $E(J) \in I(M)$. $\square$

Example 3.7. *In Example 3.2, $M = \{0, i, j, 1\}$ is a JK-algebra, which is not good, since $i^{-\sim} = 0 \neq j = i^{\sim -}$. Routine calculations show that $T = \{0, i\} \in I(M)$, after calculations, we obtain that $D(T) = \{1\} \in F(M)$, $E(T) = \{j, 1\} \notin F(M)$, since $j \in E(T)$, $j \rightsquigarrow i = j \in E(T)$, but $i \notin E(T)$.*

Example 3.8. *In Example 3.4, $M = \{0, i, j, k, 1\}$ is a JK-algebra, which is not involutive, since $i^{\sim -} = 0 \neq i$. Routine calculations show that $J = \{j, 1\} \in F(M)$, after calculations, we obtain that $D(J) = \{0, k\} \notin I(M)$ and $E(J) = \{0, i\} \in I(M)$.*

Proposition 3.9. *Let M be an involutive pseudo equality algebra. Then the following properties hold for any $s, t, w \in M$,*
(1) $s^\sim \to t = t^- \rightsquigarrow s$,
(2) $s^\sim \to (t^\sim \to w) = (s^\sim \to t)^\sim \to w$, $s^- \rightsquigarrow (t^- \rightsquigarrow w) = (s^- \rightsquigarrow t)^- \rightsquigarrow w$.

Proof. (1) By $(M18)$, we have

$$\begin{aligned} s^\sim \to t &= s^\sim \to t^{-\sim} \\ &= s^\sim \to (t^- \rightsquigarrow 0) \\ &= t^- \rightsquigarrow (s^\sim \to 0) \\ &= t^- \rightsquigarrow s^{\sim -} = t^- \rightsquigarrow s. \end{aligned}$$

(2) By (1) and (M18), we have

$$s^\sim \to (t^\sim \to w) = s^\sim \to (w^- \rightsquigarrow t)$$
$$= w^- \rightsquigarrow (s^\sim \to t)$$
$$= (s^\sim \to t)^\sim \to w.$$

Also, we can prove $s^- \rightsquigarrow (t^- \rightsquigarrow w) = (s^- \rightsquigarrow t)^- \rightsquigarrow w$. □

Suppose that M is an involutive pseudo equality algebra and $\varnothing \neq A \subseteq M$. The smallest ideal of M containing A is called the ideal generated by A and is denoted by $<A>$.

Theorem 3.10. *Let M be an involutive pseudo equality algebra and $\varnothing \neq A \subseteq M$. Then*
$<A> = \{a \in M \mid a \leq (\cdots((s_1^\sim \to s_2)^\sim \to s_3)^\sim \cdots \to s_n),$ *for some $n \in N$ and* $s_1, s_2, \cdots, s_n \in A\}$ *(*)*
$= \{a \in M \mid a \leq (\cdots((s_1^- \rightsquigarrow s_2)^- \rightsquigarrow s_3)^- \cdots \rightsquigarrow s_n),$ *for some $n \in N$ and* $s_1, s_2, \cdots, s_n \in A\}$ *(**)*

Proof. Let $S = \{a \in M \mid a \leq (\cdots((s_1^\sim \to s_2)^\sim \to s_3)^\sim \cdots \to s_n)$, for some $n \in N$ and $s_1, s_2, \cdots, s_n \in A\}$. For any $a \in A$, exist $s_1 = a \in A$ and $n = 1$, such that $s_1^\sim \to a = a^\sim \to a \geq a^{\sim -} \geq a$, so $a \in S$. Hence $A \subseteq S$. Since $0 \in S$, then $S \neq \varnothing$. Assume that $b, c \in M$, $b \leq c$ and $c \in S$. Since $c \in S$, then there exist $n \in N$ and $s_1, s_2, \cdots, s_n \in A$ such that $c \leq (\cdots((s_1^\sim \to s_2)^\sim \to s_3)^\sim \cdots \to s_n)$. It follows from $b \leq c$ that $b \leq (\cdots((s_1^\sim \to s_2)^\sim \to s_3)^\sim \cdots \to s_n)$. Thus $b \in S$. Assume that $b, c \in S$, then there exist $n \in N$ and $s_1, s_2, \cdots, s_n \in A$, $t_1, t_2, \cdots, t_n \in A$, satisfying $b \leq (\cdots((s_1^\sim \to s_2)^\sim \to s_3)^\sim \cdots \to s_n)$ and $c \leq (\cdots((t_1^\sim \to t_2)^\sim \to t_3)^\sim \cdots \to t_n)$, thus $(\cdots((s_1^\sim \to s_2)^\sim \to s_3)^\sim \cdots \to s_n)^\sim \leq b^\sim$. Hence,
$b^\sim \to c \leq (\cdots((s_1^\sim \to s_2)^\sim \to s_3)^\sim \cdots \to s_n)^\sim \to c$
$\leq (\cdots((s_1^\sim \to s_2)^\sim \to s_3)^\sim \cdots \to s_n)^\sim \to (\cdots((t_1^\sim \to t_2)^\sim \to t_3)^\sim \cdots \to t_n)$
$= (\cdots((s_1^\sim \to s_2)^\sim \to s_3)^\sim \cdots \to s_n)^\sim \to (\cdots t_1^\sim \to (t_2^\sim \cdots \to (t_{n-1}^\sim \to t_n)))$
$= ((\cdots((s_1^\sim \to s_2)^\sim \to s_3)^\sim \cdots \to s_n)^\sim \to t_1)^\sim \to (\cdots t_2^\sim \cdots \to (t_{n-1}^\sim \to t_n))$
$= \cdots = ((\cdots((s_1^\sim \to s_2)^\sim \to s_3)^\sim \cdots \to s_n)^\sim \to t_1)^\sim \cdots \to t_n$,
and thus $b^\sim \to c \in S$. Therefore, $S \in I(M)$. For any $a \in S$, exist $s_1, s_2, \cdots, s_n \in A \subseteq T$ such that $a \leq (\cdots((s_1^\sim \to s_2)^\sim \to s_3)^\sim \cdots \to s_n)$. Since $A \subseteq T$ and $T \in I(M)$, then $(\cdots((s_1^\sim \to s_2)^\sim \to s_3)^\sim \cdots \to s_n) \in T$ and $a \in T$. Hence $S \subseteq T$. Also, we can prove (**). □

Note that $(nt)_\to = t^\sim \to (t^\sim \cdots \to (t^\sim \to t))$, $(nt)_{\rightsquigarrow} = t^- \rightsquigarrow (t^- \cdots \rightsquigarrow (t^- \rightsquigarrow t))$.

Proposition 3.11. Let M be an involutive lattice pseudo equality algebra and $T \in I(M)$, $t \in M$. Then $<T \cup \{t\}> = \{a \in M \mid a \leq s_1^\sim \to (s_2^\sim \cdots \to (s_n^\sim \to (nt)_\to))$, for some $n \in N$ and $s_1, s_2, \cdots, s_n \in T\} \cup \{a \in M \mid a \leq t^\sim \to (t^\sim \cdots \to (t^\sim \to s))$, $s \in T\}$ $\quad$ (***)
$= \{a \in M \mid a \leq s_1^- \rightsquigarrow (s_2^- \cdots \rightsquigarrow (s_n^- \rightsquigarrow (nt)_\rightsquigarrow))$, for some $n \in N$ and $s_1, s_2, \cdots, s_n \in T\} \cup \{a \in M \mid a \leq t^- \rightsquigarrow (t^- \cdots \rightsquigarrow (t^- \rightsquigarrow s))$, $s \in T\}$. $\quad$ (****)

Proof. Let $S = \{a \in M \mid a \leq s_1^\sim \to (s_2^\sim \cdots \to (s_n^\sim \to (nt)_\to))$, for some $n \in N$ and $s_1, s_2, \cdots, s_n \in T\} \cup \{a \in M \mid a \leq t^\sim \to (t^\sim \cdots \to (t^\sim \to s))$, $s \in T\}$. For any $t \in M$, exist $s = t \in T$ and $a = t$, such that $t^\sim \cdots \to (t^\sim \to t) \geq t^\sim \to (t^\sim \to t) \geq t^\sim \to t \geq t$, so $t \in S$. Hence $\{t\} \subseteq S$. For any $a \in T$, exist $s_1 = a \in T$, $t = 0 \in T$ and $n = 1$, such that $s_1 \to 0 = a^{\sim\sim} \geq a$, so $a \in S$. Hence $T \subseteq S$. Therefore, we have $(T \cup \{t\}) \subseteq S$ and $0 \in S$. Now we prove that $S \in I(M)$. Assume that $b, c \in M$ with $b \leq c$ and $c \in S$. Then there exist $n \in N$, $s \in T$ and $s_1, s_2, \cdots, s_n \in T$ such that $c \leq s_1^\sim \to (s_2^\sim \cdots \to (s_n^\sim \to (nt)_\to))$ or $c \leq t^\sim \to (t^\sim \cdots \to (t^\sim \to s))$. Therefore, $b \leq c \leq s_1^\sim \to (s_2^\sim \cdots \to (s_n^\sim \to (nt)_\to))$ or $b \leq c \leq t^\sim \to (t^\sim \cdots \to (t^\sim \to s))$. Hence $b \in S$. Let $b, c \in S$, then there exist $s, k \in T$, $n, m \in N$, $s_1, s_2, \cdots, s_n \in T$ and $k_1, k_2, \cdots, k_m \in T$ satisfying $b \leq s_1^\sim \to (s_2^\sim \cdots \to (s_n^\sim \to (nt)_\to))$ or $b \leq t^\sim \to (t^\sim \cdots \to (t^\sim \to s))$ and $c \leq k_1^\sim \to (k_2^\sim \cdots \to (k_m^\sim \to (mt)_\to))$ or $c \leq t^\sim \to (t^\sim \cdots \to (t^\sim \to k))$.
Case 1, if $b \leq s_1^\sim \to (s_2^\sim \cdots \to (s_n^\sim \to (nt)_\to))$ and $c \leq k_1^\sim \to (k_2^\sim \cdots \to (k_m^\sim \to (mt)_\to))$, then by Proposition 2.3, we have
$b^\sim \to c \leq (s_1^\sim \to (s_2^\sim \cdots \to (s_n^\sim \to (nt)_\to)))^\sim \to c$
$\leq (s_1^\sim \to (s_2^\sim \cdots \to (s_n^\sim \to (nt)_\to)))^\sim \to (k_1^\sim \to (k_2^\sim \cdots \to (k_m^\sim \to (mt)_\to)))$
$= s_1^\sim \to ((s_2^\sim \cdots \to (s_n^\sim \to (nt)_\to))^\sim \to (k_1^\sim \to (k_2^\sim \cdots \to (k_m^\sim \to (mt)_\to))))$
$= \cdots = s_1^\sim \to (s_2^\sim \cdots \to (((nt)_\to)^\sim \to (k_1^\sim \to (k_2^\sim \cdots \to (k_m^\sim \to (mt)_\to)))))$.
Hence, $b^\sim \to c \in S$.
Case 2, if $b \leq s_1^\sim \to (s_2^\sim \cdots \to (s_n^\sim \to (nt)_\to))$ and $c \leq t^\sim \to (t^\sim \cdots \to (t^\sim \to k))$, then we have
$b^\sim \to c \leq (s_1^\sim \to (s_2^\sim \cdots \to (s_n^\sim \to (nt)_\to)))^\sim \to t^\sim \to (t^\sim \cdots \to (t^\sim \to k))$
$= s_1^\sim \to ((s_2^\sim \cdots \to (s_n^\sim \to (nt)_\to))^\sim \to (t^\sim \cdots \to (t^\sim \to k)))$
$= \cdots = s_1^\sim \to (s_2^\sim \cdots \to (((nt)_\to)^\sim \to (t^\sim \cdots \to (t^\sim \to k))))$.
Hence, $b^\sim \to c \in S$.
Other cases are analogous to the Case 1 and Case 2. Hence, $S \in I(M)$.

Now, let $B \in I(M)$ with $(T \cup \{t\}) \subseteq B$, for each $b \in S$. Then there exist $n \in N$, $s \in T$ and $s_1, s_2, \cdots, s_n \in T$ such that $b \leq s_1^\sim \to (s_2^\sim \cdots \to (s_n^\sim \to (nt)_\to))$ or $b \leq t^\sim \to (t^\sim \cdots \to (t^\sim \to s))$. Since $B \in I(M)$ and $(T \cup \{t\}) \subseteq B$, then there exist $n \in N$, $s \in B$ and $s_1, s_2, \cdots, s_n \in B$ such that $b \leq s_1^\sim \to (s_2^\sim \cdots \to (s_n^\sim \to (nt)_\to)) \in B$ and $t^\sim \to (t^\sim \cdots \to (t^\sim \to s)) \in B$. Hence, $b \in B$ and thus $S \subseteq B$. Similarly, we can prove (****). □

Proposition 3.12. *Let M be an involutive lattice pseudo equality algebra and $T \in I(M)$, $d, g \in M$. Then $<T \cup \{d\}> \cap <T \cup \{g\}> = <T \cup \{d \wedge g\}>$.*

Proof. The proof is similar to Proposition 3.11. □

A subset $\theta \subseteq M \times M$ is called a congruence of M, if it is an equivalence relation and for all $s_1, t_1, s_2, t_2 \in M$ such that $(s_1, t_1), (s_2, t_2) \in \theta$ the following hold,
(CG_1) $(s_1 \wedge s_2, t_1 \wedge t_2) \in \theta$,
(CG_2) $(s_1 \sim s_2, t_1 \sim t_2) \in \theta$,
(CG_2) $(s_1 \backsim s_2, t_1 \backsim t_2) \in \theta$.
Denote the set of all congruences of M by $Con(M)$.

Theorem 3.13. *Let M be a bounded involutive pseudo equality algebra and $T \in I(M)$. The binary relation $\frown_T$ on M is defined by*

$$s \frown_T t \text{ iff } (s^- \rightsquigarrow t^-)^\sim \in T \text{ and } (t^- \rightsquigarrow s^-)^\sim \in T, (s^\sim \to t^\sim)^- \in T \text{ and } (t^\sim \to s^\sim)^- \in T.$$

Then $\frown_T$ is an equivalence relation on M.

Proof. For every $s \in M$ and by $(M13)$, $(s^- \rightsquigarrow s^-)^\sim = 0 \in T$, $(s^\sim \to s^\sim)^- = 0 \in T$, then $s \frown_T s$. Thus $\frown_T$ is reflexivity. Since the definition of $\frown_T$, we have $\frown_T$ is symmetry. Let $s, t, w \in M$, $s \frown_T t$ and $t \frown_T w$. We have $(s^- \rightsquigarrow t^-)^\sim \in T$, $(t^- \rightsquigarrow s^-)^\sim \in T$ and $(t^- \rightsquigarrow w^-)^\sim \in T$, $(w^- \rightsquigarrow t^-)^\sim \in T$. By $(M16)$ and $(M25)$,

$$((s^- \rightsquigarrow t^-)^{\sim -} \rightsquigarrow (s^- \rightsquigarrow w^-)^{\sim -})^\sim = ((s^- \rightsquigarrow t^-) \rightsquigarrow (s^- \rightsquigarrow w^-))^\sim$$
$$\leq (t^- \rightsquigarrow w^-)^\sim \in T.$$

Since $T \in I(M)$, we have $(s^- \rightsquigarrow t^-)^\sim \in T$. By $(MT5)$, $(s^- \rightsquigarrow w^-)^\sim \in T$. Similarly, $(w^- \rightsquigarrow s^-)^\sim \in T$. By the similarly way, we have $(s^\sim \to t^\sim)^- \in T$ and $(t^\sim \to s^\sim)^- \in T$. Hence, $\frown_T$ is transitive. So $\frown_T$ is a equivalence relation on M. □

Moerover, by Proposition 2.10, $(s^- \rightsquigarrow s^{-\sim -})^\sim = 0 \in T$ and $(s^{-\sim -} \rightsquigarrow s^-)^\sim = 0 \in T$, $(s^\sim \to s^{\sim -\sim})^- = 0 \in T$ and $(s^{\sim -\sim} \to s^\sim)^- = 0 \in T$. Hence, $s \frown_T s^{-\sim}$. Similarly, $s \frown_T s^{\sim -}$.

The following example shows that the equivalence relation $\frown_T$ defined in Theorem 3.14 is not a congruence relation.

Example 3.14. *In Example 3.3, we take $T = \{0, i\}$. Routine calculations show that $\smile_{\{0,i\}} = \{(0,0), (0,i), (i,0), (j,1), (1,j), (i,i), (j,j), (1,1)\}$. After calculations, we can see that $\smile_{\{0,i\}}$ is not a congruence relation. Note that $(0,i), (j,1) \in \smile_{\{0,i\}}$, $(0 \smile j, i \smile 1) = (1,i) \notin \smile_{\{0,i\}}$.*

Theorem 3.15. *Let M be a symmetric involutive pseudo equality algebra and $T \in I(M)$. The binary relation $\approx_T$ on M is defined by*

$$s \approx_T t \text{ iff } (s^- \smile t^-)^\sim \text{ and } (t^- \smile s^-)^\sim \in T, (s^\sim \sim t^\sim)^- \text{ and } (t^\sim \sim s^\sim)^- \in T.$$

Then $\approx_T$ is a congruence relation on M.

Proof. For every $s \in M$, $(s^- \smile s^-)^\sim = 0 \in T$, $(s^\sim \sim s^\sim)^- = 0 \in T$, $s \approx_T s$. Thus $\approx_T$ is reflexivity. Since the definition of $\approx_T$, we have $\approx_T$ is symmetry. Let $s, t, w \in M$, $s \approx_T t$ and $t \approx_T w$, we get that $(s^- \smile t^-)^\sim, (t^- \smile s^-)^\sim \in T$ and $(t^- \smile w^-)^\sim, (w^- \smile t^-)^\sim \in T$. By $(M6), (M7), (M8)$ and $(M25)$,

$$\begin{aligned}
((s^- \smile t^-)^{\sim -} \rightsquigarrow (s^- \smile w^-)^{\sim -})^\sim &\leq ((s^- \smile t^-)^{\sim -} \smile (s^- \smile w^-)^{\sim -})^\sim \\
&= ((s^- \smile t^-) \smile (s^- \smile w^-))^\sim \\
&\leq (t^- \smile w^-)^\sim \in T.
\end{aligned}$$

Since $(t^- \smile w^-)^\sim \in T$, we get that $((s^- \smile t^-)^{\sim -} \rightsquigarrow (s^- \smile w^-)^{\sim -})^\sim \in T$. Since $(s^- \smile t^-)^\sim \in T$ and $(MT5)$, we have $(s^- \smile w^-)^\sim \in T$. Similarly,

$$\begin{aligned}
((t^- \smile s^-)^{\sim -} \rightsquigarrow (w^- \smile s^-)^{\sim -})^\sim &\leq ((t^- \smile s^-)^{\sim -} \smile (w^- \smile s^-)^{\sim -})^\sim \\
&= ((t^- \smile s^-) \smile (w^- \smile s^-))^\sim \\
&= ((w^- \smile s^-) \sim (t^- \smile s^-))^\sim \\
&\leq (w^- \smile t^-)^\sim \in T.
\end{aligned}$$

By the similarly way, $(s^\sim \sim w^\sim)^-$ and $(w^\sim \sim s^\sim)^- \in T$. Thus $\approx_T$ is transitive. So we have $\approx_T$ is an equivalence relation on M. Assume that $s, t, w \in M$, $s \approx_T t$, then $(s^- \smile t^-)^\sim, (t^- \smile s^-)^\sim \in T$. By $(M5), (M6)$ and $(M7)$,

$$\begin{aligned}
((s \wedge w)^- \smile (t \wedge w)^-)^\sim &\leq ((s \wedge w) \sim (t \wedge w))^\sim \\
&\leq (s \sim t)^\sim \\
&= (s^{-\sim} \sim t^{-\sim})^\sim \\
&\leq (s^- \smile t^-)^\sim \in T,
\end{aligned}$$

$$\begin{aligned}
((t \wedge w)^- \smile (s \wedge w)^-)^\sim &\leq ((t \wedge w) \sim (s \wedge w))^\sim \\
&\leq (t \sim s)^\sim \\
&= (t^{-\sim} \sim s^{-\sim})^\sim \\
&\leq (t^- \smile s^-)^\sim \in T.
\end{aligned}$$

Hence, $((s \wedge w)^- \backsmile (t \wedge w)^-)^\sim, ((t \wedge w)^- \backsmile (s \wedge w)^-)^\sim \in T$. By the similarly way, we have $((s \wedge w)^\sim \backsmile (t \wedge w)^\sim)^-, ((t \wedge w)^\sim \backsmile (s \wedge w)^\sim)^- \in T$. Thus $(s \wedge w) \approx_T (t \wedge w)$. Moreover, by $(M6)$ and $(M7)$,

$$\begin{aligned}
((s \sim w)^- \backsmile (t \sim w)^-)^\sim &\leq ((s \sim w) \sim (t \sim w))^\sim \\
&\leq (s \sim t)^\sim \\
&= (s^{-\sim} \sim t^{-\sim})^\sim \\
&\leq (s^- \backsmile t^-)^\sim \in T,
\end{aligned}$$

$$\begin{aligned}
((t \sim w)^- \backsmile (s \sim w)^-)^\sim &\leq ((t \sim w) \sim (s \sim w))^\sim \\
&\leq (t \sim s)^\sim \\
&= (t^{-\sim} \sim s^{-\sim})^\sim \\
&\leq (t^- \backsmile s^-)^\sim \in T.
\end{aligned}$$

Hence, $((s \sim w)^- \backsmile (t \sim w)^-)^\sim, ((t \sim w)^- \backsmile (s \sim w)^-)^\sim \in T$. By the similarly way, we have $((s \backsmile w)^\sim \sim (t \backsmile w)^\sim)^-, ((t \backsmile w)^\sim \sim (s \backsmile w)^\sim)^- \in T$. Thus $(s \sim w) \approx_T (t \sim w)$.
By $(M6)$ and $(M7)$,

$$\begin{aligned}
((s \backsmile w)^- \backsmile (t \backsmile w)^-)^\sim &\leq ((s \backsmile w) \backsmile (t \backsmile w))^\sim \\
&\leq (s \backsmile t)^\sim \\
&= (t \sim s)^\sim \\
&= (t^{-\sim} \sim s^{-\sim})^\sim \\
&\leq (t^- \backsmile s^-)^\sim \in T,
\end{aligned}$$

Hence, $((s \backsmile w)^- \backsmile (t \backsmile w)^-)^\sim, ((t \backsmile w)^- \backsmile (s \backsmile w)^-)^\sim \in T$. By the similarly way, we have $((s \backsmile w)^\sim \backsmile (t \backsmile w)^\sim)^-, ((t \backsmile w)^\sim \backsmile (s \backsmile w)^\sim)^- \in T$. Thus $(s \backsmile w) \approx_T (t \backsmile w)$.
Therefore, $\approx_T$ is a congruence relation on M. $\square$

Example 3.16. *[6] Let $M = \{0, i, j, 1\}$ in which the Hasse diagram and the operations $\backsmile$ and $\sim$ on M given as follows,*

Table 7: Cayley table for the binary operation "$\sim$"

$\sim$	0	i	j	1
0	1	j	i	0
i	1	1	i	i
j	1	j	1	j
1	1	1	1	1

341

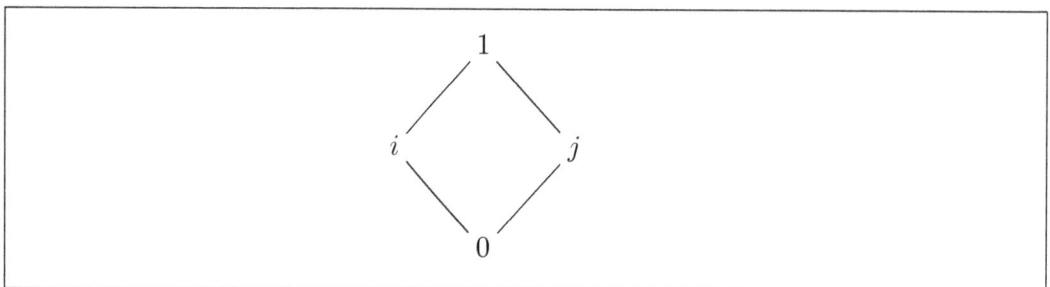

Figure 3: Hasse Diagram of M

Table 8: Cayley table for the binary operation "$\backsim$"

$\backsim$	0	i	j	1
0	1	1	1	1
i	j	1	i	i
j	i	i	1	1
1	0	i	j	1

After calculations, we can observe that M is a JK-algebra, which is involutive, not symmetric, since $i \backsim j = i \neq j = j \sim i$. $T_1 = \{0, i\}$ is an ideal of M. Routine calculations show that $\approx_{T_1} = \{(0,0), (0,i), (i,0), (j,1), (1,j), (i,i), (j,j), (1,1)\}$ is an equivalence relation, which is not a congruence relation on M. Note that $(0, i), (j, j) \in \approx_{T_1}$, $(0 \backsim j, i \backsim j) = (1, i) \notin \approx_{T_1}$.

Let $(M, \wedge, \sim, \backsim, 1)$ be a symmetric involutive pseudo equality algebra and $T \in I(M)$. Define $M/T := \{[s] \mid s \in M\}$, $[s] = \{t \in M \mid s \approx_T t\}$. For any $s, t \in M$, the operations $\wedge_T, \backsim_T, \sim_T$ are defined by,

$$[s] \wedge_T [t] = [s \wedge t], [s] \backsim_T [t] = [s \backsim t], [s] \sim_T [t] = [s \sim t].$$

Theorem 3.17. Let $(M, \wedge, \sim, \backsim, 1)$ be a symmetric involutive pseudo equality algebra and $T \in I(M)$. Then $(M/T, \wedge_T, \sim_T, \backsim_T, 1/T)$ is a symmetric involutive pseudo equality algebra.

Proof. The proof is obvious. □

Proposition 3.18. Let M be a bounded pseudo equality algebra and $T \in I(M)$. Then the following hold,
(i) if M is involutive and 0 is invariant, then $[0]_T = \{s \in M \mid s \approx_T 0\} \in I(M)$,
(ii) if σ is a congruence on M, then $[0]_\sigma = \{(s, 0) \in \sigma\} \in I(M)$.

Proof. (i)
$$[0]_T = \{s \in M \mid s \approx_T 0\}$$
$$= \{(s \in M) \mid (s^- \backsim 0^-)^\sim, (0^- \backsim s^-)^\sim \in T\}.$$

By (M3), $(0^- \backsim s^-)^\sim = (1 \backsim s^-)^\sim$. Since M is involutive, $s^{-\sim} = s \in T$. By (M8), $s^- \backsim 0^- = s^- \backsim 1 \leq s^- \leadsto 1 = 1$. Since M is bounded, $0 \leq s$. By (M21), (M23) and Proposition 2.10, we have

$$s^- \backsim 0^- = s^- \backsim 1 = 1 \sim s^- = 1 \sim (s \to 0) = 1 \sim (0 \sim s) \geq s \sim 0 = 1.$$

Hence, $(s^- \backsim 0^-)^\sim = 1^\sim = 0 \in T$.
Thus
$$[0]_T = \{s \in m \mid 0 \in T, s \in T\}$$
$$= \{s \in M \mid s \in T\}$$
$$= T.$$

Therefore, $[0]_T \in I(M)$.

(ii) Assume that $s, t \in M$, $s \leq t$ and $t \in [0]_\sigma$, we have $(t, 0) \in \sigma$. Since σ is a congruence, we have $(s \wedge t, s \wedge 0) \in \sigma$ and $(s, 0) \in \sigma$. Hence, $s \in [0]_\sigma$. Let $s, t \in [0]_\sigma$. Since $(0,0) \in \sigma$, $(s \backsim 0, 0 \backsim 0) \in \sigma$. We can conclude that $(s^\sim, 1) \in \sigma$, $(s^\sim \wedge t, 1 \wedge t) \in \sigma$ and thus $(s^\sim \wedge t, t) \in \sigma$. Since $(s^\sim, 1) \in \sigma$ and $(s^\sim \wedge t, t) \in \sigma$, we can conclude that $((s^\sim \wedge t) \sim s^\sim, t \sim 1) \in \sigma$. Hence $((s^\sim \wedge t) \sim s^\sim, t) \in \sigma$. Since $(t, 0) \in \sigma$, we have $((s^\sim \wedge t) \sim s^\sim, 0) \in \sigma$. Thus $s^\sim \to t = (s^\sim \wedge t) \sim s^\sim \in [0]_\sigma$. Therefore, $[0]_\sigma \in I(M)$. □

Example 3.19. *In Example 3.3, $M = \{0, i, j, 1\}$ is a JK-algebra, which is involutive, the element 0 is not invariant, since $0 \backsim i = j \neq 1$. Routine calculations show that $[0]_{T_4} = \{0, i, j\} \notin I(M)$.*

Theorem 3.20. *Let M be a symmeyric involutive pseudo equality algebra. There is one-to-one correspondence between $I(M)$ and $Con(M)$.*

Proof. Define $\varphi : Con(M) \to I(M)$, $\varphi(\sigma) = [0]_\sigma$. By Proposition 3.18, we have $\varphi(\sigma) \in I(M)$. If $s \in [0]_{\sigma_1}$, then $(s, 0) \in \sigma_1$. Since $\sigma_1 = \sigma_2$, we have $(s, 0) \in \sigma_2$, and so $s \in [0]_{\sigma_2}$. Therefore, $[0]_{\sigma_1} \subseteq [0]_{\sigma_2}$. By the similar way, we conclude that $[0]_{\sigma_2} \subseteq [0]_{\sigma_1}$. Thus $[0]_{\sigma_1} = [0]_{\sigma_2}$, i.e., $\varphi(\sigma_1) = \varphi(\sigma_2)$. Therefore, φ is a well defined mapping. Let $(s, t) \in \sigma$. Since σ is a congruence, $(s^-, t^-) \in \sigma$ and $(t^-, s^-) \in \sigma$, i.e., $(s^- \backsim t^-, t^- \backsim t^-) \in \sigma$. Thus $(s^- \backsim t^-, 1) \in \sigma$. Hence, $((s^- \backsim t^-)^\sim, 1^\sim) \in \sigma$. Therefore, we have $(s^- \backsim t^-)^\sim \in [0]_\sigma$. Similarly, $(t^- \backsim s^-)^\sim \in [0]_\sigma$. By Proposition 3.18, we have $s \approx_{[0]_\sigma} t$, i.e., $(s, t) \in \approx_{[0]_\sigma}$. Hence, we obtain that $\sigma \subseteq \approx_{[0]_\sigma}$. Conversely, suppose that $(s, t) \in \approx_{[0]_\sigma}$, i.e., $(s^- \backsim t^-)^\sim \in [0]_\sigma$ and $(t^- \backsim s^-)^\sim \in [0]_\sigma$. Hence, $((s^- \backsim t^-)^\sim, 0) \in \sigma$. Then $((s^- \backsim t^-)^{\sim\sim}, 0^-) \in \sigma$. Since M is symmyric, we get

343

that $(s^- \backsim t^-, 1) = (t^- \sim s^-, 1) \in \sigma$. Furthermore, since σ is a congruence, we have $((t^- \sim s^-) \backsim t^-, 1 \backsim t^-) \in \sigma$ and $(((t^- \sim s^-) \backsim t^-) \wedge s^-, t^- \wedge s^-) \in \sigma$. Since $s^- \leq (t^- \sim s^-) \backsim t^-$, we have $(s^-, t^- \wedge s^-) \in \sigma$. By the similar way, $(t^-, t^- \wedge s^-) \in \sigma$. Since σ is transitive, then $(s^{-\sim}, t^{-\sim}) \in \sigma$. Hence, $(s,t) \in \sigma$ and $\approx_{[0]_\sigma} \subseteq \sigma$. Therefore, $\approx_{[0]_\sigma} = \sigma$. Let $\sigma_1, \sigma_2 \in Con(M)$ such that $\varphi(\sigma_1) = \varphi(\sigma_2)$. Then $[0]_{\sigma_1} = [0]_{\sigma_2}$, we have $\approx_{[0]_{\sigma_1}} = \approx_{[0]_{\sigma_2}}$. Therefore, $\sigma_1 = \sigma_2$. Hence, φ is an one-to-one mapping. If $T \in I(M)$, then $s \in T$ iff $(s^- \backsim 0^-)^\sim = 0 \in T$, $(0^- \backsim s^-)^\sim = s^{-\sim} = s \in T$ iff $s \approx_T 0$ iff $(s,0) \in \approx_T$ iff $s \in [0]_{\approx_T}$, so $T = [0]_{\approx_T}$. By Theorem 3.16, we have φ is an onto mapping. Hence, we can get that φ is an onto correspondence between $I(M)$ and $Con(M)$. □

Definition 3.21. *Let $(M_1, \wedge_1, \sim_1, \backsim_1, 1_{M_1})$ and $(M_2, \wedge_2, \sim_2, \backsim_2, 1_{M_2})$ be two pseudo equality algebras. Then a mapping $f : M_1 \to M_2$ is called a pseudo equality homomorphism, if for all $s, t \in M_1$,*
(i) $f(s \wedge_1 t) = f(s) \wedge_2 f(t)$,
(ii) $f(s \sim_1 t) = f(s) \sim_2 f(t)$,
(iii) $f(s \backsim_1 t) = f(s) \backsim_2 f(t)$.

Lemma 3.22. *Let $(M_1, \wedge_1, \sim_1, \backsim_1, 1_{M_1})$ and $(M_2, \wedge_2, \sim_2, \backsim_2, 1_{M_2})$ be two pseudo equality algebras and $f : M_1 \to M_2$ be a pseudo equality homomorphism. Then for all $s, t \in M_1$,*
(i) $f(s \to_1 t) = f(s) \to_2 f(t)$, $f(s \rightsquigarrow_1 t) = f(s) \rightsquigarrow_2 f(t)$,
(ii) $f(1_{M_1}) = 1_{M_2}$,
(iii) $f(s^-) = (f(s))^-$, $f(s^\sim) = (f(s))^\sim$.

Proof. The proof is obvious. □

Proposition 3.23. *Let $f : M_1 \to M_2$ be a homomorphism of pseudo equality algebras. The following hold,*
(1) if $T \in I(M_2)$, then $f^{-1}(T) \in I(M_1)$,
(2) if f is surjective, and $T \in I(M_1)$, then $f(T) \in I(M_2)$,
(3) if $kerf = \{s \in M_1 \mid f(s) = 0\}$, then $kerf \subset I(M_1)$.

Proof. (1) Let $T \in I(M_2)$. Since $f(0) = 0 \in T$, we have $0 \in f^{-1}(T)$. Suppose that $s \leq t$ and $t \in f^{-1}(T)$. Then $f(t) \in T$. Since $s \to t = 1$ and f is a homomorphism, we obtain that $f(s) \leq f(t)$. Thus $f(s) \in T$. So $s \in f^{-1}(T)$. Let $s, t \in f^{-1}(T)$, then $f(s), f(t) \in T$. Since $T \in I(M_2)$, $f(s^\sim \to t) = (f(s))^\sim \to f(t) \in T$. Thus $s^\sim \to t \in f^{-1}(T)$ and so $f^{-1}(T) \in I(M_1)$.
(2) Let $T \in I(M_1)$. We have $0 \in f(T)$. Let $s \leq t$ and $t \in f(T)$, then there exists $a \in T$ such that $f(a) = t$. Since $s = f(b) \leq f(a) = t$, we have $f(1_{M_1}) =$

$1_{M_2} = f(b) \to f(a) = f(b \to a)$. Thus $b \leq a$. Moreover, since $T \in I(M_1)$ and $a \in T$, we have $b \in T$. Thus $s = f(b) \in f(T)$. Let $s, t \in f(T)$, then there exist $a, b \in I$ such that $f(a) = s$ and $f(b) = t$. Since $T \in I(M_1)$, $a^\sim \to b \in T$. Thus $s^\sim \to t = (f(a))^\sim \to f(b) \in f(T)$. Hence, $f(T) \in I(M_2)$.
(3) This is the result of (1). □

Theorem 3.24. *Let M_1, M_2 be two involutive pseudo equality algebras and $f : M_1 \to M_2$ be a pseudo equality homomorphism. Then $M_1/\ker f \cong \mathrm{Im} f$.*

Proof. The proof is obvious. □

4 Prime ideals on pseudo equality algebras

In this section, we will introduce the definitions of prime ideals and maximal ideals and discuss related properties.

Definition 4.1. *Let $(M, \wedge, \sim, \backsim, 1)$ be a bounded pseudo equality algebra and $M \neq R \in I(M)$. The R is called a prime ideal, if it fulfills: for any $s, t \in M$, $(s \to t)^\sim \in R$ or $(t \to s)^\sim \in R$ and $(s \rightsquigarrow t)^- \in R$ or $(t \rightsquigarrow s)^- \in R$.*

Denote the set of all prime ideals by $P(M)$.

Example 4.2. *[7] In Example 3.2, $T_1 = \{0\}$, $T_2 = \{0, i\}$ and M are the ideals. After calculations, we can see that T_1 and T_2 are the prime ideals of M.*

Example 4.3. *In Example 3.3, after calculations, we can see that $T_2 = \{0, i\}$ and $T_3 = \{0, j\}$ are the prime ideals of M. $T_1 = \{0\} \notin P(M)$, since $(i \to j)^\sim = j^\sim = i \notin T_1$, $(j \to i)^\sim = i^\sim = j \notin T_1$.*

Example 4.4. *In Example 3.4, after calculations, we can see that $T_2 = \{0, i\} \in P(M)$. $I_1 = \{0\} \notin P(M)$, since $(j \to k)^\sim = i^\sim = j \notin T_1$, $(k \to j)^\sim = j^\sim = i \notin T_1$.*

Theorem 4.5. *Let M be a bounded prelinear pseudo equality algebra and $M \neq R \in I(M)$. Then $R \in P(M)$ iff for every $s, t \in M$, $s \wedge t \in R$ implies $s \in R$ or $t \in R$.*

Proof. ($\Rightarrow$) Let $R \in P(M)$. For $s, t \in M$, $(s \to t)^\sim \in R$ or $(t \to s)^\sim \in R$ and $(s \rightsquigarrow t)^- \in R$ or $(t \rightsquigarrow s)^- \in R$.
Case 1, Let $s \wedge t \in R$ and $(s \to t)^\sim, (s \rightsquigarrow t)^- \in R$, for $s, t \in M$. Then by $(M16)$ and $(M20)$,
$$((s \wedge t)^- \rightsquigarrow s^-)^\sim \leq (s \to (s \wedge t))^\sim$$
$$= (s \to t)^\sim,$$

$$((s \wedge t)^\sim \to s^\sim)^- \leq (s \rightsquigarrow (s \wedge t))^-$$
$$= (s \rightsquigarrow t)^-.$$

Since $R \in P(M)$, $s \wedge t \in R$ and $(s \to t)^\sim, (s \rightsquigarrow t)^- \in R$, by Theorem 3.5, we have $s \in R$.

Case 2, the proof is similar to Case 1, if $s \wedge t \in R$ and $(t \to s)^\sim, (t \rightsquigarrow s)^- \in R$, then we have $t \in R$.

Case 3 and Case 4 are analogous to the Case 1 and Case 2. If $s \wedge t \in R$ and $(s \to t)^\sim \in R$, $(t \rightsquigarrow s)^- \in R$, then $s \in R$ and $t \in R$. If $s \wedge t \in R$ and $(t \to s)^\sim \in P$, $(s \rightsquigarrow t)^- \in R$, then $s \in R$ and $t \in R$.

($\Leftarrow$) Since M is prelinear, we have $(s \to t) \vee (t \to s) = 1 = (s \rightsquigarrow t) \vee (t \rightsquigarrow s)$, for every $s, t \in M$. By Proposition 2.10, $(s \to t)^\sim \wedge (t \to s)^\sim = 0 \in R$ and $(s \rightsquigarrow t)^- \wedge (t \rightsquigarrow s)^- = 0 \in R$. Hence, $(s \to t)^\sim \in R$ or $(t \to s)^\sim \in R$ and $(s \rightsquigarrow t)^- \in R$ or $(t \rightsquigarrow s)^- \in R$. Therefore, $R \in P(M)$. □

Theorem 4.6. *Let M be a symmetric involutive pseudo equality algebra and $R \in I(M)$. Then $R \in P(M)$ iff M/R is a chain.*

Proof. ($\Rightarrow$) Assume that $s, t \in M$, $(s \to t)^\sim \in R$ or $(t \to s)^\sim \in R$ and $(s \rightsquigarrow t)^- \in R$ or $(t \rightsquigarrow s)^- \in R$. By Proposition 2.10 and Proposition 3.18, $((s \to t)^{\sim -} \rightsquigarrow 0^-)^\sim = 0, (0^- \rightsquigarrow (s \to t)^{\sim -})^\sim = (s \to t)^\sim \in R$ or $((t \to s)^{\sim -} \rightsquigarrow 0^-)^\sim = 0, (0^- \rightsquigarrow (t \to s)^{\sim -})^\sim = (t \to s)^\sim \in R$ and $((s \rightsquigarrow t)^{-\sim} \to 0^\sim)^- = (s \rightsquigarrow t)^-, (0^\sim \to (s \rightsquigarrow t)^{-\sim})^- = 0 \in R$ or $((t \rightsquigarrow s)^{-\sim} \to 0^\sim)^- = (t \rightsquigarrow s), (0^\sim \to (t \rightsquigarrow s)^{-\sim})^- = 0 \in R$. Thus $[(s \to t)^\sim] = [0]$ or $[(t \to s)^\sim] = [0]$ and $[(s \rightsquigarrow t)^-] = [0]$ or $[(t \rightsquigarrow s)^-] = [0]$. Hence, $[(s \to t)^{\sim -}] = [0^-]$ or $[(t \to s)^{\sim -}] = [0^-]$ and $[(s \rightsquigarrow t)^{-\sim}] = [0^\sim]$ or $[(t \rightsquigarrow s)^{-\sim}] = [0^\sim]$. Since M is involutive, we have $[s \to t] = [1]$ or $[t \to s] = [1]$ and $[s \rightsquigarrow t] = [1]$ or $[t \rightsquigarrow s] = [1]$. Thus $[s] \to [t] = [1]$ or $[t] \to [s] = [1]$ and $[s] \rightsquigarrow [t] = [1]$ or $[t] \rightsquigarrow [s] = [1]$. Therefore, $[s] \leq [t]$ or $[t] \leq [s]$. We can conclude M/R is a chain.

($\Leftarrow$) Let M/R be a chain. Then $[s] \leq [t]$ or $[t] \leq [s]$. Thus $[s] \to [t] = [1]$ or $[t] \to [s] = [1]$ and $[s] \rightsquigarrow [t] = [1]$ or $[t] \rightsquigarrow [s] = [1]$. Hence, $((s \to t)^\sim \rightsquigarrow 1^\sim)^-$, $(1^\sim \rightsquigarrow (s \to t)^\sim)^- \in R$ or $((t \to s)^\sim \rightsquigarrow 1^\sim)^-$, $(1^\sim \rightsquigarrow (t \to s)^\sim)^- \in R$ and $((s \rightsquigarrow t)^- \to 1^-)^\sim$, $(1^- \to (s \rightsquigarrow t)^-)^\sim \in R$ or $((t \rightsquigarrow s)^- \to 1^-)^\sim$, $(1^- \to (t \rightsquigarrow s)^-)^\sim \in R$. Thus $((s \to t)^\sim)^{\sim -} \in R$ or $((t \to s)^\sim)^{\sim -} \in R$ and $((s \rightsquigarrow t)^-)^{-\sim} \in R$ or $((t \rightsquigarrow s)^-)^{-\sim} \in R$. Since M is involutive, we have $(s \to t)^\sim \in R$ or $(t \to s)^\sim \in R$ and $(s \rightsquigarrow t)^- \in R$ or $(t \rightsquigarrow s)^- \in R$. Therefore, $R \in P(M)$. □

Proposition 4.7. *Let M be a bounded pseudo equality algebra. Then the following hold,*

(1) if M is good and $R \in P(M)$, then $D(R)$ and $E(R)$ are the prime filters of M,

(2) *if M is involutive and K be a prime filter of M, then $D(K)$ and $E(K)$ are the prime ideals of M.*

Proof. The proof is analogous to that of the Proposition 3.6. □

Proposition 4.8. *Let $f : M_1 \to M_2$ be a homomorphism of pseudo equality algebras. The following hold,*
(1) *if $R \in P(M_2)$, then $f^{-1}(R) \in P(M_1)$,*
(2) *if f is surjective and $R \in P(M_1)$, then $f(R) \in M_2$.*

Proof. (1) Let $R \in P(M_2)$. By Proposition 3.23, $f^{-1}(R)$ is an ideal of M_1. Let $s, t \in P(M_1)$, we have $f(s), f(t) \in P(M_2)$. Since $R \in P(M_2)$, then $(f(s) \to f(t))^\sim \in R$ or $(f(t) \to f(s))^\sim \in R$ and $(f(s) \rightsquigarrow f(t))^- \in R$ or $(f(t) \rightsquigarrow f(s))^- \in R$. By Lemma 3.22, $f((s \to t)^\sim) \in R$ or $f((t \to s)^\sim) \in R$ and $f((s \rightsquigarrow t)^-) \in R$ or $f((t \rightsquigarrow s)^-) \in R$. Thus $(s \to t)^\sim \in f^{-1}(R)$ or $(t \to s)^\sim \in f^{-1}(R)$ and $(s \rightsquigarrow t)^- \in f^{-1}(R)$ or $(t \rightsquigarrow s)^- \in f^{-1}(R)$. Therefore, $f^{-1}(R) \in P(M_1)$.
(2) Let $R \in P(M_1)$. By Proposition 3.23, $f(R)$ is an ideal of M_2. Let $s, t \in f(R)$, then there exist $a, b \in R$ such that $f(a) = s$ and $f(b) = t$. Since $R \in P(M_1)$, $(a \to b)^\sim \in R$ or $(b \to a)^\sim \in R$ and $(a \rightsquigarrow b)^- \in R$ or $(b \rightsquigarrow a)^- \in R$. By Lemma 3.22, $(s \to t)^\sim = (f(a) \to f(b))^\sim = f((a \to b)^\sim) \in f(R)$ or $(t \to s)^\sim = (f(b) \to f(a))^\sim = f((b \to a)^\sim) \in f(R)$ and $(s \rightsquigarrow t)^- = (f(a) \rightsquigarrow f(b))^- = f((a \rightsquigarrow b)^-) \in f(R)$ or $(t \rightsquigarrow s)^- = (f(b) \rightsquigarrow f(a))^- = f((b \rightsquigarrow a)^-) \in f(R)$. Therefore, $f(R) \in P(M_2)$. □

Proposition 4.9. *Let M be a bounded prelinear pseudo equality algebra and $M \neq R \in I(M)$. Then $R \in P(M)$ iff for any $T, Q \in I(M)$, $T \cap Q \subseteq R$ implies $T \subseteq R$ or $Q \subseteq R$.*

Proof. ($\Rightarrow$) Let $R \in P(M)$, $T, Q \in I(M)$ with $T \cap Q \subseteq R$. If $T \nsubseteq R$ and $Q \nsubseteq R$, then there exist $s \in T \setminus R$ and $t \in Q \setminus R$. Since $s \wedge t \leq s, t$, then $s \wedge t \in T$ and $s \wedge t \in Q$. Hence, $s \wedge t \in T \cap Q \subseteq R$. Thus $s \wedge t \in R$. By Theorem 4.5, $s \in R$ or $t \in R$, which is contradiction. Therefore, $T \subseteq R$ or $Q \subseteq R$.
($\Leftarrow$) Let $M \neq R \in I(M)$ and $s \wedge t \in R$ for each $s, t \in M$. If $s, t \notin R$, then by Proposition 3.12, $< R \cup \{s\} > \cap < R \cup \{t\} > = < R \cup \{s \wedge t\} > = R$. Thus $< R \cup \{s\} > \subseteq R$ or $< R \cup \{t\} > \subseteq R$. Thus $s \in R$ or $t \in R$, which is contradiction. Hence, $R \in P(M)$. □

Proposition 4.10. *Let M be a pseudo equality algebra and $M \neq R \in I(M)$. If M is a chain, then $R \in P(M)$.*

Proof. Let $R \in I(M)$. Then for $s, t \in M$, we have $s \leq t$ or $t \leq s$. Suppose that $s \leq t$. Hence, $s \to t = 1$, $s \rightsquigarrow t = 1$. Thus $(s \to t)^\sim = 0 \in R$, $(s \rightsquigarrow t)^- = 0 \in R$.

By the similar way, if $t \leq s$, then we have $(t \to s)^\sim = 0 \in R$, $(t \rightsquigarrow s)^- = 0 \in R$. Therefore, $R \in P(M)$.

The following example shows that the reverse is not true, if $M \neq R \in P(M)$, M is not necessarily a chain. □

Example 4.11. *[21] Let $M = \{0, i, j, k, 1\}$ in which the Hasse diagram and the operations $\smile$ and $\sim$ on M given as follows,*

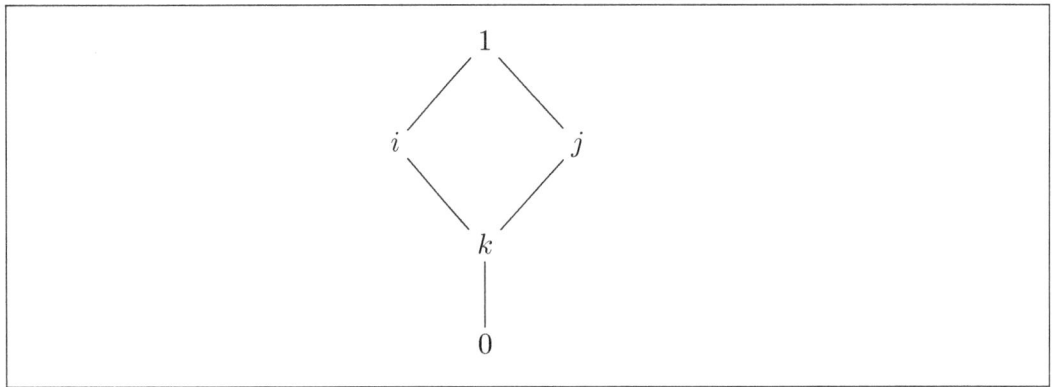

Figure 4: Hasse Diagram of M

Table 9: Cayley table for the binary operation "$\sim$"

$\sim$	0	k	i	j	1
0	1	j	j	0	0
k	1	1	j	i	k
i	1	1	1	i	i
j	1	1	j	1	j
1	1	1	1	1	1

Table 10: Cayley table for the binary operation "$\smile$"

$\smile$	0	k	i	j	1
0	1	1	1	1	1
k	i	1	1	1	1
i	0	j	1	j	1
j	i	i	i	1	1
1	0	k	i	j	1

By routine calculations, we get that M is a JK-algebra, $T_1 = \{0\}$, $T_2 = M$ are the ideals of M. We can prove that T_1 is a prime ideal of M.

Definition 4.12. *Let $M \neq G \in I(M)$. If no proper ideal strictly contains G, then G is called a maximal ideal.*

Denote the set of all maximal ideals by $Max(M)$.

Example 4.13. *In Example 3.3, after calculations, we can see that $T_2 = \{0, i\}$ and $T_3 = \{0, j\}$ are the maximal ideals of M.*

Example 4.14. *In Example 3.4, after calculations, we can see that $T_2 = \{0, i\}$ is a maximal ideal of M.*

Theorem 4.15. *Let M be an involutive lattice pseudo equality algebra. Then every maximal ideal of M is a prime ideal.*

Proof. Clearly $M \neq G \in I(M)$. Suppose $s, t \in M$ with $s \wedge t \in G$. If $s \notin G$, then $G \subsetneq < G \cup \{s\} >$. Since G is a maximal ideal, we have $< G \cup \{s\} > = M$. By the similary way. If $t \notin G$, $< G \cup \{t\} > = M$. By Proposition 3.12, we have $M = < G \cup \{s\} > \cap < G \cup \{t\} > = < G \cup \{s \wedge t\} > = G$, which is a contradiction. Thus $G \in P(M)$. □

Note that, in general, a prime ideal is not necessarily a maximal ideal as shown by the following example.

Example 4.16. *[21] Let $M = \{0, i, j, m, n, e, f, g, h, 1\}$ in which the Hasse diagram and the operations $\curvearrowright$ and $\sim$ on M given as follows,*

Table 11: Cayley table for the binary operation "$\sim$"

$\sim$	0	i	j	m	n	e	f	g	h	1
0	1	h	g	f	e	n	j	m	i	0
i	1	1	g	f	e	n	j	m	i	i
j	1	1	1	f	f	n	j	n	j	j
m	1	1	g	1	g	n	n	m	m	m
n	1	1	1	1	1	n	n	n	n	n
e	1	1	1	1	1	1	g	f	e	e
f	1	1	1	1	1	1	1	f	f	f
g	1	1	1	1	1	1	g	1	g	g
h	1	1	1	1	1	1	1	1	1	h
1	1	1	1	1	1	1	1	1	1	1

Table 12: Cayley table for the binary operation "⤳"

⤳	0	i	j	m	n	e	f	g	h	1
0	1	1	1	1	1	1	1	1	1	1
i	h	1	1	1	1	1	1	1	1	1
j	f	f	1	f	1	1	1	1	1	1
m	g	g	g	1	1	1	1	1	1	1
n	e	e	g	f	1	1	1	1	1	1
e	n	n	n	n	n	1	1	1	1	1
f	m	m	n	n	n	g	1	g	1	1
g	j	j	j	n	n	f	f	1	1	1
h	i	i	j	m	n	e	f	g	1	1
1	0	i	j	m	n	e	f	g	h	1

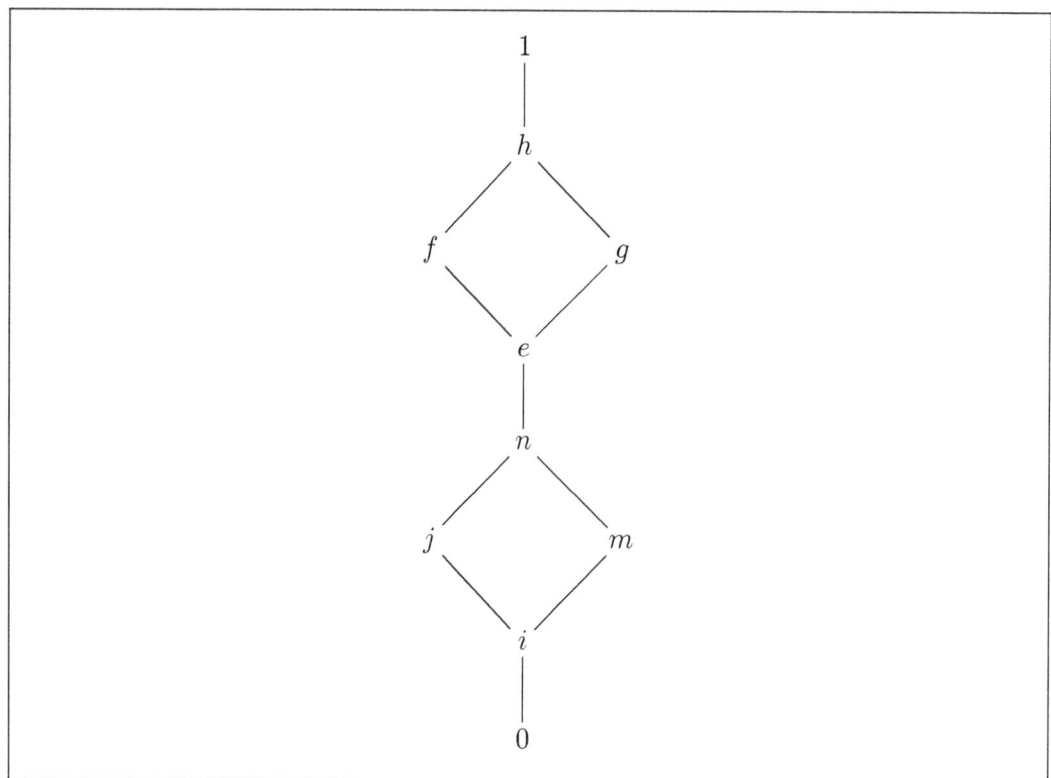

Figure 5: Hasse Diagram of M

By routine calculations, we get that M is an involutive lattice pseudo equality algebra. Routine calculations show that $T_1 = \{0\}$, $T_2 = \{0, i, j\}$, $T_3 = \{0, i, m\}$, $T_4 = \{0, i, j, m, n\}$ and $T_5 = M$ are the ideals of M. After calculations, we obtain that T_2 is a prime ideal of M, but it is not a maximal ideal of M.

Example 4.17. [21] Let $M = \{0, i, j, m, n, 1\}$ in which the Hasse diagram and the operations $\backsim$ and $\sim$ on M given as follows,

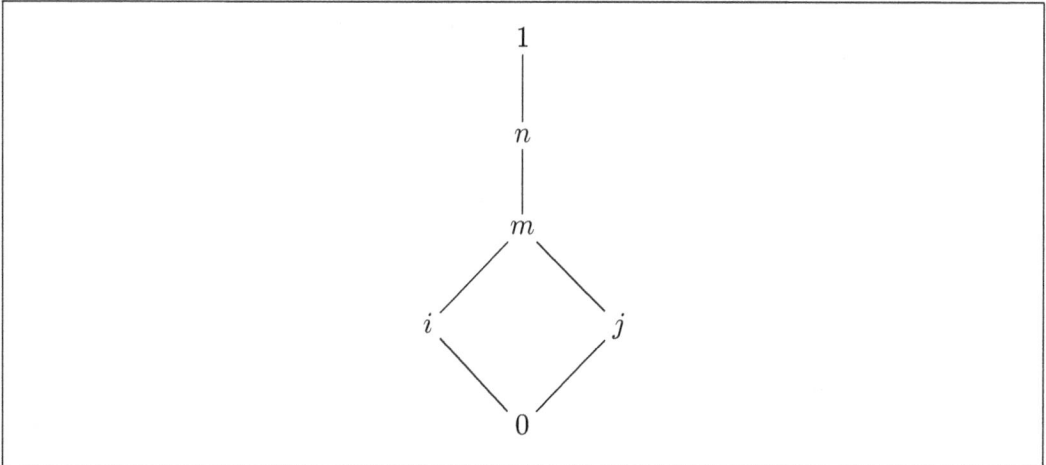

Figure 6: Hasse Diagram of M

Table 13: Cayley table for the binary operation "$\sim$"

$\sim$	0	i	j	m	n	1
0	1	n	n	n	0	0
i	1	1	n	n	i	i
j	1	n	1	n	j	j
m	1	1	1	1	m	m
n	1	1	1	1	1	n
1	1	1	1	1	1	1

Table 14: Cayley table for the binary operation "$\backsim$"

$\rightsquigarrow$	0	i	j	m	n	1
0	1	1	1	1	1	1
i	m	1	m	1	1	1
j	m	m	1	1	1	1
m	m	m	m	1	1	1
n	m	m	m	m	1	1
1	0	i	j	m	n	1

By routine calculations, we get that M is a JK-algebra, which is not involutive, since $i^{\sim -} = n \neq i$. Routine calculations show that $T_1 = \{0\}$, $T_2 = M$ are the ideals of M. $T_1 = \{0\}$ is a maximal ideal of M. After calculations, we obtain that T_1 is not a prime ideal of M, since $(i \rightarrow j)^\sim = n^\sim = m \notin T_1$, $(j \rightarrow i)^\sim = n^\sim = m \notin T_1$.

Proposition 4.18. *Let M be a pseudo equality algebra and $M \neq G \in I(M)$. Then $G \in Max(M)$ iff $< G \cup \{t\} > = M$ for any $t \in M$ and $t \notin G$.*

Proof. ($\Rightarrow$) If $t \notin G$, then $G \subsetneq < G \cup \{t\} > \subseteq M$. Since $G \in Max(M)$, we have $< G \cup \{t\} > = M$.
($\Leftarrow$) Let W be an ideal of M such that $G \subsetneq W \subseteq M$. Then there exists $t \in W$ with $t \notin G$. Hence, we have $M = < G \cup \{t\} > \subseteq W$, and so $W = M$. Therefore, $G \in Max(M)$. $\square$

5 Conclusion

In this paper, we first defined ideals on pseudo equality algebras with the operations of $\sim \rightarrow$, $- \rightsquigarrow$, and gave several examples of pseudo equality algebras. We provided an equivalent characterization of ideals on good pseudo equality algebras and discussed the relationships between ideals and filters. Next, we presented the generation formula of ideals on involutive pseudo equality algebras. Then we induced congruence relations by ideals and constructed quotient pseudo equality algebras. In particular, we proved that the equivalence class of 0 with respect to the ideal T can only induce an ideal of M if the pseudo equality algebra is involutive, $0 \in M$ and 0 is invariant. We introduced the concept of prime ideals and maximal ideals and showed that if a pseudo equality algebra is a chain, then all its ideals are prime ideals. We also gave an example to demonstrate that the reverse is not true.

In our future work, we plan to use prime ideals to investigate the topological space on pseudo equality algebras and use topological structures to study the geometric properties of pseudo equality algebras.

References

[1] R.A. Borzooei, B.G. Saffar, *States on EQ-algebras*, Journal of Intelligent and Fuzzy Systems, 29(1) (2015): 209-221.

[2] C.C. Chang, *Algebraic analysis of many valued logics*, Transactions of the American Mathematic Society, 88 (1958): 467-490.

[3] L.C. Ciungu, *Commutative pseudo-equality algebras*, Soft Comput, 20 (2016): 1-16.

[4] L.C. Ciungu, *On pseudo equality algebras*, Mathematical Logic, 53 (2014): 561-570.

[5] W.J. Chen, W.A. Dudek, *Ideals and congruences in quasi-pseudo-MV algebras*, Soft Computing, 22 (2018): 3879-3889.

[6] X.Y. Cheng, X.L. Xin, P.F. He, *Generalized state maps and states on pseudo equality algebras*, Open Mathematics, 16(01) (2018): 133-148.

[7] A. Dvurecčenskij, O. Zahiri, *Pseudo equality algebras: revision*, Soft Comput, 20(6) (2016): 2091-2101.

[8] A. Dvurecčenskij, O. Zahiri, *Weak pseudo EMV-algebras. I: basic properties*, Journal of Applied Logic, 8(10) (2021): 2365-2399.

[9] M. El-Zekey, V. Novák, R. Mesiar, *On good EQ-algebras.* Fuzzy Sets and Systems, 178 (2011): 1-23.

[10] G. Georgescu, A. Iorgulescu, *Pseudo-MV algebras*, Multiple-Valued Logic, 6(1-2) (2001): 95-135.

[11] P. Hájek, *Metamathematics of fuzzy logic*, Dordrecht, Kluwer Academic Publishers, (1998).

[12] S. Jenei, *Equality algebras*, Studia Logica, 100(6) (2012): 1201-1209.

[13] S. Jenei, Kóródi, *Pseudo equality algebras*, Mathematical Logic, 52 (2013): 469-481.

[14] M.K. Liu, X.L. Xin, *Filter theory of pseudo equality algebras*, Journal of Intelligent & Fuzzy Systems, 39 (2020): 475-487.

[15] V. Novák, B.De Baets, *EQ-algebras*, Fuzzy Sets and Systems, 160 (2009): 2956-2978.

[16] A. Paad, *Ideals in bounded equality algebras*, Filomat, 33(7) (2019): 2113-2123.

[17] J.Q. Shi, X.L. Xin, *Ideal theory on EQ-algebras*, AIMS Mathematics, 6(11) (2021): 11686-11707.

[18] M. Ward, R.P. Dilworth, *Residuated lattices*, Transactions of the American Mathematic Society, 45 (1939): 335-354.

[19] F. Xie, H. Liu, *Ideals in pseudo-hoop algebras*, Journal of Algebraic Hyperstructures and Logical Algebras, 1(4) (2020): 39-53.

[20] X.Y. Cheng, X.L. Xin, P.F. He, *Generalized state maps and states on pseudo equality algebras*, Open Mathematics, 16(1) (2018): 133-148.

[21] M.P. Zhao, Z.H. Chen, Y.J. Xie, *The relations between some types of pseudo equality algebras*, Fuzzy Systems and Mathematics, 35(04) (2021): 1-14.

Some Types of Weak Hyper Filters in Hyper BE-algebras

XIAOYUN CHENG*
School of Science, Xi'an Aeronautical Institute, China
chengxiaoyun2004@163.com

XIAOGUANG LI
School of Science, Xi'an Aeronautical Institute, China
1225122250163.com

XIAOLI GAO
School of Mathematics and Information Science, Xianyang Normal University, China
yagxl0203@163.com

Abstract

In this paper, some types of weak hyper filters in hyper BE-algebras are introduced and studied including positive implicative weak hyper filters, implicative weak hyper filters and obstinate weak hyper filters. The relationships between (positive) implicative weak hyper filters and weak hyper filters, and also obstinate weak hyper filters and maximal weak hyper filters, positive implicative hyper filters, are discussed respectively. Moreover, the equivalent characterizations of these weak hyper filters are given, and the corresponding conditions are found.

keyword: hyper BE-algebra; weak hyper filter; (positive) implicative weak hyper filter; obstinate weak hyper filter

---

The authors are very grateful to the editors and the referees for their valuable comments and suggestions which are helpful for improving this paper. This research is partially supported by the Natural Science Basic Research Plan in Shaanxi Province of China (No. 2022JM-014,2022JQ-068), the Scientific Research Program of Shaanxi Provincial Department of Education (No. 21JK0963)and PHD Research Start-up Foundation of Xi'an Aeronautical Institute.
*Corresponding author.

1 Introduction

The theory of hyper algebras was first proposed by F. Marty [15] in 1934 at the 8th Scandinavian Congress of Mathematicians. It refers to an algebraic system with hyper operations is an extension of the original algebra. The so-called hyper operation refers to the operation that the combination of two elements is a set rather than an element. The hyper structure has important practical significance, for example, pea hybridization, REDOX reaction and other practical problems in genetics can be abstracted into the models of hyper structures with hyper operations [9, 10].

As we all know, fuzzy logic is an important tool to deal with uncertain information, and non-classical logical algebra is the corresponding algebraic semantics of fuzzy logic, which plays an important role in the research of fuzzy logic. As a generalization of non-classical logical algebras, many hyper algebraic structures have been proposed and studied such as hyper BCK-algebras [12, 13, 17], hyper K-algebras [11, 24, 25], hyper BF-algebras [16], hyper residuated lattices [1, 26], hyper EQ-algebras [4, 7] and hyper equality algebras [3, 6], etc. At present, the hyper algebraic theory has been widely applied in pure mathematics and applied mathematics [8]. As a generalization of ideals and filters in logical algebras, hyper ideals and hyper filters are important substructures of hyper algebras which play important roles in studying the theory of hyper algebras. R.A. Borzooei et al. [2] introduced the notions of some types of positive implicative hyper BCK-ideals which are showed different. After that the relationship between these notions and (strong, weak) hyper BCK-ideals was investigated. A. Borumand Saeid et al. [20, 21] defined and investigate (weak) implicative, obstinate and maximal hyper K-ideals in hyper K-algebras. Then they state and prove some theorems which determine the relationship between these notions and the other hyper K-ideals. R.A. Borzooei et al. in [1] and [3] respectively studied the hyper filter theory of hyper residuated lattices and hyper equality algebras, focusing on the equivalent characterization of (positive) implicative hyper filters. Y.W. Yang et al. [22, 23] studied two kinds of fuzzy weak hyper deductive systems of hyper residuated lattices and hyper equality algebras respectively, and studied the falling shadow theory on hyper residuated lattices through fuzzy (positive) implicative hyper deductive systems. As a generalization of BE-algebra [14, 19], A. Radfar et al. [18] introduced the notion of hyper BE-algebras which is a generalization of dual hyper K-algebras and dual hyper BCK-algebras. They gave some related properties and defined (weak) hyper filters in hyper BE-algebras. Also, they pointed out that hyper filters are weak hyper filters, but the converse is not true.

Based on the above analysis, as a dual generalization of weak hyper BCK-ideals and weak hyper K-ideals in hyper BCK-algebras and hyper K-algebras, this present paper intends to consider several kinds of weak hyper filters in hyper BE-algebras so as to further characterize and master the structure of hyper BE-algebras. At the meantime, it may lay a theoretical foundation for the study of hyper BE-logic.

This paper is organized as follows: in Section 2, we review and give some basic concepts and results in hyper BE-algebras. In Section 3 and Section 4, we introduce (positive) implicative weak hyper filters and mainly give some characterizations of them. In Section 5 we introduce obstinate weak hyper filters in hyper BE-algebras and deliver some characterizations, and moreover we discuss the relationships between obstinate weak hyper filters and other types of weak hyper filters.

2 Preliminaries

In this section, we recollect and propose some definitions and results about hyper BE-algebras which will be used in the following.

Definition 2.1. *[18] Let H be a nonempty set and $\circ : H \times H \to P^*(H)$ be a hyper operation. Then $(H, \circ, 1)$ is called a hyper BE-algebra provided it satisfies the following axioms:*

(HBE1) $x \ll 1$ and $x \ll x$;
(HBE2) $x \circ (y \circ z) = y \circ (x \circ z)$;
(HBE3) $x \in 1 \circ x$;
(HBE4) $1 \ll x$ implies $x = 1$,

for all $x, y \in H$, where the relation $\ll$ is defined by $x \ll y \Leftrightarrow 1 \in x \circ y$. For any two nonempty subsets A and B of H, $A \ll B$ means that there exist $a \in A, b \in B$ such that $a \ll b$.

Notice that in any hyper BE-algebra, $A \circ B = \bigcup_{a \in A, b \in B} a \circ b$ and $A \leq B$ means for any $a \in A$, there exists $b \in B$ such that $a \ll b$. In what follows, by H denote a hyper BE-algebra $(H, \circ, 1)$, unless otherwise specified.

Proposition 2.2. *[5, 18] For any $x, y \in H, A, B \subseteq H$, we have:*

(1) $A \circ (B \circ C) = B \circ (A \circ C)$;
(2) $A \subseteq 1 \circ A, 1 \in A \circ 1, 1 \in A \circ A$;
(3) $x \leq y \circ x, A \leq B \circ A$;
(4) $A \ll B$ iff $1 \in A \circ B$;
(5) $1 \in A$ and $A \leq B$ imply $1 \in B$;
(6) $1 \ll A$ implies $1 \in A$.

Definition 2.3. *[18] A hyper BE-algebra H is said to be a*
  (1) *R-hyper BE-algebra, if $1 \circ x = \{x\}$ for all $x \in H$;*
  (2) *C-hyper BE-algebra, if $x \circ 1 = \{1\}$ for all $x \in H$;*
  (3) *D-hyper BE-algebra), if $x \circ x = \{1\}$ for all $x \in H$;*
  (4) *CD-hyper BE-algebra, if H is both a C-hyper BE-algebra and a D-hyper BE-algebra.*

Definition 2.4. *[18] By a hyper subalgebra of H we mean a nonempty subset S of H which satisfies $x \circ y \subseteq S$ whence $x, y \in S$.*

Definition 2.5. *[18] A subset F containing 1 of H is said to be a*
  (1) *weak hyper filter if $x \circ y \subseteq F$ and $x \in F$ imply $y \in F$ for all $x, y \in H$;*
  (2) *hyper filter if $x \circ y \cap F \neq \emptyset$ and $x \in F$ imply $y \in F$ for all $x, y \in H$.*

According to [18] every hyper filter F of H is a weak hyper filter and moreover it satisfies the condition (F):
  (F) $x \in F$ and $x \ll y$ imply $y \in F$ for all $x, y \in H$.

Definition 2.6. *[5] A nonempty subset S of H is said to be $\circ$-reflexive if $x \circ y \cap S \neq \emptyset$ implies $x \circ y \subseteq S$ for all $x, y \in H$.*

Proposition 2.7. *Let S be a $\circ$-reflexive nonempty subset of H.*
  (1) *If $A \subseteq S$, then $1 \circ A \subseteq S$;*
  (2) *If S satisfies the condition (F), then S is a hyper subalgebra of H.*

Proof. (1) and (2) can be obtained immediately by $A \subseteq 1 \circ A$ and $x \leq y \circ x$, respectively. □

Proposition 2.8. *Let F be a $\circ$-reflexive weak hyper filter of H. Then*
  (1) *F satisfies the condition (F);*
  (2) *F is a hyper subalgebra of H;*
  (3) *$A \cap F \neq \emptyset$ and $A \leq B$ imply $B \cap F \neq \emptyset$ for any $A, B \subseteq H$.*

Proof. (1) It is trivial.
  (2) By (1) and Proposition 2.7.
  (3) By (1) and the definition of $A \leq B$. □

3 Positive implicative weak hyper filters

In this section, we introduce positive implicative weak hyper filters in hyper BE-algebras, and investigate the relationship between weak hyper filters and positive implicative weak hyper filters. Moreover we deliver characterizations of positive implicative weak hyper filters.

Definition 3.1. *A nonempty subset F of H is said to be a positive implicative weak hyper filter if it satisfies:*
  (1) $1 \in F$;
  (2) $x \circ ((y \circ z) \circ y) \subseteq F$ and $x \in F$ imply $y \in F$ for all $x, y, z \in H$.

Example 3.2. (1) *Consider $H = \{1, a, b\}$ with the operation $\circ$ defined by the table:*

$\circ$	1	a	b
1	$\{1\}$	$\{a, b\}$	$\{b\}$
a	$\{1\}$	$\{1, a, b\}$	$\{b\}$
b	$\{1, b\}$	$\{1, a, b\}$	$\{1, a, b\}$

Then $(H, \circ, 1)$ is a hyper BE-algebras [18]. One can check that $F = \{1, a\}$ and $G = \{1, b\}$ are two positive implicative weak hyper filters of H.
  (2) *Consider $H = \{1, a, b, c, d\}$ with the operation $\circ$ defined by the table:*

$\circ$	1	a	b	c	d
1	$\{1, a, b, c\}$	$\{a\}$	$\{b\}$	$\{c\}$	$\{d\}$
a	$\{1, a, b, c\}$	$\{1, a, b, c\}$	$\{a, b, c\}$	$\{1, a, b, c\}$	$\{b\}$
b	$\{1, a, b, c\}$	$\{a, b, c\}$	$\{1, a, b, c\}$	$\{a, b, c\}$	$\{a\}$
c	$\{1, a, b, c\}$	$\{a, b, c\}$	$\{1, a, b, c\}$	$\{1, a, b, c\}$	$\{a, c\}$
d	$\{1, a, b, c\}$	$\{1, a, b, c\}$	$\{1, a, b, c\}$	$\{1, a, b, c\}$	$\{1\}$

Then routine to check that $(H, \circ, 1)$ is a hyper BE-algebra and $F = \{1, a, b, c\}$ is not a positive implicative weak hyper filter of H since $a \in F$, $a \circ ((d \circ 1) \circ d) = \{1, a, b, c\} \subseteq F$, but $d \notin F$.

Not every positive implicative weak hyper filter of a hyper BE-algebra is a weak hyper filter in general.

Example 3.3. *Consider $H = \{1, a, b\}$ with the operation $\circ$ given by the table:*

$\circ$	1	a	b
1	$\{1\}$	$\{a\}$	$\{b\}$
a	$\{1, b\}$	$\{1, a, b\}$	$\{1, a\}$
b	$\{1, a, b\}$	$\{a\}$	$\{1, a, b\}$

Then $(H, \circ, 1)$ is a hyper BE-algebras [18]. Easy to verify that $F = \{1, a\}$ is a positive implicative weak hyper filters of H, but it is not a weak hyper filter since $a \circ b = \{1, a\} \subseteq F$ and $a \in F$ while $b \notin F$.

In what follows, we provide conditions that a positive implicative weak hyper filter of a hyper BE-algebra becomes a weak hyper filter.

Proposition 3.4. *Suppose that F is a $\circ$-reflexive hyper subalgebra of H. If F is a positive implicative weak hyper filter, then F is a weak hyper filter.*

Proof. Let $x \in F$ and $x \circ y \subseteq F$. It follows from $1 \in y \circ 1$ that $y \circ 1 \cap F \neq \emptyset$. Since F is $\circ$-reflexive, we have $y \circ 1 \subseteq F$. Combining that $x \circ y \subseteq F$ and F is a hyper subalgebra, it yields that $x \circ ((y \circ 1) \circ y) = (y \circ 1) \circ (x \circ y) \subseteq F$. Noticing that $x \in F$ and F is a positive implicative weak hyper filter, we can conclude that $y \in F$. Therefore F is a weak hyper filter. $\square$

Example 3.5. *Consider $H = \{1, a, b, c\}$ in which $\circ$ is defined by the table:*

$\circ$	1	a	b	c
1	$\{1,a\}$	$\{a\}$	$\{b\}$	$\{c\}$
a	$\{1,a\}$	$\{1,a\}$	$\{b\}$	$\{c\}$
b	$\{1,a\}$	$\{1,a\}$	$\{1,a\}$	$\{c\}$
c	$\{1,a\}$	$\{1,a\}$	$\{1,a\}$	$\{1,a\}$

One can check that $(H, \circ, 1)$ is a hyper BE-algebra and $F = \{1, a, b\}$ is a $\circ$-reflexive hyper subalgebra of H. Moreover F is both a weak hyper filter and a positive implicative weak hyper filter.

Remark 3.6. (1) *The conditions of Proposition 3.4 are not necessary in general. In fact, in Example 3.2 (1) $F = \{1, a\}$ is not $\circ$-reflexive since $1 \circ a = \{a, b\} \cap F \neq \emptyset$ while $1 \circ a \not\subseteq F$, and also it is not a hyper subalgebra since $1 \circ a = \{a, b\} \not\subseteq F$. However F is both a weak hyper filter and a positive implicative weak hyper filter.*

(2) *The condition of the $\circ$-reflexivity from Proposition 3.4 is not necessary in general. Consider the hyper BE-algebra from Example 3.2 (1). It can be verified that $F = \{1\}$ is a hyper subalgebra but it is not $\circ$-reflexive since $a \circ b = \{1, a\} \cap F \neq \emptyset$ while $a \circ b \not\subseteq F$. However F is both a weak hyper filter and a positive implicative weak hyper filter.*

(3) *The condition of the hyper subalgebra in Proposition 3.4 is not necessary in general. Suppose $H = \{1, a, b, c, d\}$ and the operation $\circ$ is given by the table:*

$\circ$	1	a	b	c	d
1	$\{1,d\}$	$\{a\}$	$\{b\}$	$\{c\}$	$\{d\}$
a	$\{1,d\}$	$\{1,d\}$	$\{b\}$	$\{c\}$	$\{d\}$
b	$\{1,d\}$	$\{a\}$	$\{1,d\}$	$\{c\}$	$\{d\}$
c	$\{1,d\}$	$\{a\}$	$\{b\}$	$\{1,d\}$	$\{d\}$
d	$\{1,d\}$	$\{a\}$	$\{b\}$	$\{c\}$	$\{1,d\}$

Then $(H, \circ, 1)$ is a hyper BE-algebra [11]. One can verify that $F = \{1, a\}$ is not a hyper subalgebra since $a \circ a = \{1, d\} \not\subseteq F$. Moreover F is both a weak hyper filter and a positive implicative weak hyper filter.

Proposition 3.7. *Suppose that H is a D-hyper (C-hyper) BE-algebra and F is a $\circ$-reflexive nonempty subset of H. If F is a positive implicative weak hyper filter of H, then F is a weak hyper filter.*

Proof. Let $x \in F$ and $x \circ y \subseteq F$. Since $y \in 1 \circ y$ then $x \circ y \subseteq x \circ (1 \circ y) = 1 \circ (x \circ y)$. Considering that $x \circ y \subseteq F$ and F is $\circ$-reflexive, it follows from (1) of Proposition 2.7 that $1 \circ (x \circ y) \subseteq F$. That is, $x \circ ((y \circ y) \circ y) = (y \circ y) \circ (x \circ y) \subseteq F$ $(x \circ ((y \circ 1) \circ y) = (y \circ 1) \circ (x \circ y) \subseteq F)$. Again since $x \in F$ and F is a positive implicative weak hyper filter, we can conclude that $y \in F$. It implies that F is a weak hyper filter. $\square$

Example 3.8. *Suppose $H = \{a, b, 1\}$ and the operation $\circ$ is given as follows:*

$\circ$	1	a	b
1	$\{1\}$	$\{a\}$	$\{b\}$
a	$\{1\}$	$\{1\}$	$\{b\}$
b	$\{1\}$	$\{1, a\}$	$\{1\}$

Then $(H, \circ, 1)$ is a CD-hyper BE-algebras [18] and $F = \{1, a\}$ is a $\circ$-reflexive subset of H. It is not difficult to verify that F is both a weak hyper filter and a positive implicative weak hyper filter.

Remark 3.9. (1) *The conditions of Proposition 3.7 are not necessary in general. In fact, in Example 3.2 (1), H is neither a C-hyper BE-algebra nor a D-hyper BE-algebra, and moreover $G = \{1, b\}$ is not $\circ$-reflexive since $b \circ a = \{1, a, b\} \cap G \neq \emptyset$ can't imply $b \circ a \subseteq G$. However it can easily calculate that $G = \{1, b\}$ is both a weak hyper filter and a positive implicative weak hyper filter.*

(2) *The condition of the $\circ$-reflexivity in Proposition 3.7 is not necessary in general. Consider $H = \{a, b, 1\}$ with the operation $\circ$ given as follows:*

$\circ$	1	a	b
1	$\{1\}$	$\{a, b\}$	$\{b\}$
a	$\{1\}$	$\{1, a\}$	$\{1, b\}$
b	$\{1\}$	$\{1, a, b\}$	$\{1\}$

Then $(H, \circ, 1)$ is a C-hyper BE-algebras [18]. One can check that $F = \{1, a\}$ is both a weak hyper filter and a positive implicative weak hyper filters, but F is not $\circ$-reflexive since $a \circ b = \{1, b\} \cap F \neq \emptyset$ while $a \circ b \not\subseteq F$.

(3) *The condition of the C-hyper (D-hyper) BE-algebra in Proposition 3.7 is not necessary in general. Consider Remark 3.6 (3), H is neither a C-hyper BE-algebra nor a D-hyper BE-algebra. It is easily verified that $F = \{1, d\}$ is $\circ$-reflexive, and moreover F is both a weak hyper filter and a positive implicative weak hyper filter.*

In general, not every weak hyper filter of a hyper BE-algebra is a positive implicative weak hyper filter. Let us see the following example.

Example 3.10. *Consider $H = \{1, a, b, c\}$ and the operation $\circ$ is given by the following table:*

$\circ$	1	a	b	c
1	$\{1\}$	$\{a\}$	$\{b\}$	$\{c\}$
a	$\{1\}$	$\{1\}$	$\{a\}$	$\{b,c\}$
b	$\{1\}$	$\{1\}$	$\{1\}$	$\{1\}$
c	$\{1\}$	$\{1\}$	$\{a\}$	$\{1,b,c\}$

Then $(H, \circ, 1)$ is a hyper BE-algebra [18]. It can be verified that $F = \{1, a\}$ is a weak hyper filter, however it is not a positive implicative weak hyper filter since $a \circ ((b \circ 1) \circ b) = \{a\} \subseteq F$ and $a \in F$ while $b \notin F$.

The converse of Proposition 3.4 and Proposition 3.7 are not true in general and see the following example.

Example 3.11. *(1) Consider the hyper BE-algebra from Example 3.5. One can check that $F = \{1, a\}$ is a $\circ$-reflexive hyper subalgebra and moreover it is a weak hyper filter, however, it is not a positive implicative weak hyper filter since $1 \circ ((b \circ c) \circ b) = \{1, a\} \subseteq F$ and $1 \in F$ while $b \notin F$.*

(2) Suppose $H = \{1, a, b, c\}$ and the operation $\circ$ is given as the following table:

$\circ$	1	a	b	c
1	$\{1\}$	$\{a\}$	$\{b\}$	$\{c\}$
a	$\{1\}$	$\{1\}$	$\{1\}$	$\{1\}$
b	$\{1\}$	$\{a\}$	$\{1,b\}$	$\{c\}$
c	$\{1\}$	$\{a\}$	$\{1,b\}$	$\{1,b\}$

Then $(H, \circ, 1)$ is a C-hyper BE-algebra [11]. It is routine to verify that $F = \{1, b\}$ is a $\circ$-reflexive subset and moreover it is a weak hyper filter. However, F is not a positive implicative weak hyper filter since $1 \circ ((c \circ a) \circ c) = \{1\} \subseteq F$ and $1 \in F$ while $c \notin F$.

In the following, we give characterizations of positive implicative weak hyper filters in hyper BE-algebras. In the meanwhile, we find the conditions that weak hyper filters of hyper BE-algebras become positive implicative weak hyper filters.

Theorem 3.12. *Let F be a $\circ$-reflexive weak hyper filter of H. Then the following are equivalent:*

(1) F is a positive implicative weak hyper filter;
(2) $(x \circ y) \circ x \subseteq F$ implies $x \in F$ for all $x, y \in H$;
(3) $z \in F$ and $(x \circ y) \circ (z \circ x) \subseteq F$ imply $x \in F$ for all $x, y \in H$.

Proof. (1) $\Rightarrow$ (2) Assume (1) holds and $(x \circ y) \circ x \subseteq F$ for any $x, y \in H$. Since $(x \circ y) \circ x \subseteq 1 \circ ((x \circ y) \circ x)$, $(x \circ y) \circ x \subseteq F$ and F is $\circ$-reflexive, it follows from (1) of Proposition 2.7 that $1 \circ ((x \circ y) \circ x) \subseteq F$. Considering that $1 \in F$ and F is a positive implicative weak hyper filter, we can obtain $x \in F$, which shows (2).

(2) $\Rightarrow$ (3) Assume that (2) holds and $z \in F, (x \circ y) \circ (z \circ x) \subseteq F$ for any $x, y \in H$. Then $z \circ ((x \circ y) \circ x) = (x \circ y) \circ (z \circ x) \subseteq F$. Since $z \in F$ and F is a weak hyper filter, we have $(x \circ y) \circ x \subseteq F$ and thus using (2) it follows that $x \in F$, which shows (3).

(3) $\Rightarrow$ (1) Assume that (3) holds and $x \in F, x \circ ((y \circ z) \circ y) \subseteq F$ for any $x, y \in H$. Then $(y \circ z) \circ (x \circ y) = x \circ ((y \circ z) \circ y) \subseteq F$. Since $x \in F$ and hence from (3) we get $y \in F$. It proves that F is a positive implicative weak hyper filter. $\square$

Corollary 3.13. *Let F be a weak hyper filter of a R-hyper BE-algebra H. Then the following are equivalent:*
(1) F is a positive implicative weak hyper filter;
(2) $(x \circ y) \circ x \subseteq F$ implies $x \in F$ for all $x, y \in H$;
(3) $z \in F$ and $(x \circ y) \circ (z \circ x) \subseteq F$ imply $x \in F$ for all $x, y \in H$.

Proof. According to Theorem 3.12, (2) $\Rightarrow$ (3) and (3) $\Rightarrow$ (1) are showed.

(1) $\Rightarrow$ (2) Assume (1) holds and $(x \circ y) \circ x \subseteq F$ for any $x, y \in H$. Since H is a R-hyper BE-algebra, it follows that $1 \circ ((x \circ y) \circ x) = (x \circ y) \circ x \subseteq F$. Considering that $1 \in F$ and F is a positive implicative weak hyper filter, we can obtain $x \in F$, which shows (2). $\square$

4 Implicative weak hyper filters

In this section, we introduce implicative weak hyper filters in hyper BE-algebras, and investigate the relationship between weak hyper filters and implicative weak hyper filters in hyper BE-algebras. Moreover we give a characterization of implicative weak hyper filters.

Definition 4.1. *A nonempty subset F of H is said to be an implicative weak hyper filter of H if it satisfies:*
(1) $1 \in F$;
(2) $x \circ (y \circ z) \subseteq F$ and $x \circ y \subseteq F$ imply $x \circ z \subseteq F$ for all $x, y, z \in H$.

Example 4.2. (1) *Consider the hyper BE-algebra H from (1) of Example 3.2. It is easy to verify that $F = \{1, a\}$ is an implicative weak hyper filter of H;*

(2) *Consider the hyper BE-algebra H from (2) of Example 3.2. One can calculate that $F = \{1, a, b, c\}$ is not an implicative weak hyper filter of H since $1 \circ (a \circ d) = 1 \circ b = \{b\} \subseteq F$ and $1 \circ a = \{a\} \subseteq F$ but $1 \circ d = \{d\} \not\subseteq F$.*

An implicative weak hyper filter of a hyper BE-algebra may not be a weak hyper filter and see the following example.

Example 4.3. *Suppose $H = \{a, b, 1\}$ in which the operation $\circ$ is given below:*

$\circ$	1	a	b
1	$\{1\}$	$\{a, b\}$	$\{b\}$
a	$\{1, b\}$	$\{1\}$	$\{1\}$
b	$\{1, b\}$	$\{1\}$	$\{1\}$

Then $(H, \circ, 1)$ is a hyper BE-algebras [18]. One can check that $F = \{1, a\}$ is an implicative weak hyper filter, but it is not a weak hyper filter since $a \circ b = \{1\} \subseteq F$ and $a \in F$ while $b \notin F$.

In what follows, we give some conditions that an implicative weak hyper filter of a hyper BE-algebra becomes a weak hyper filter.

Proposition 4.4. *Suppose that F is a $\circ$-reflexive nonempty subset of H. If F is an implicative weak hyper filter of H, then F is a weak hyper filter.*

Proof. Let $x, y \in H$ such that $x \in F$, $x \circ y \subseteq F$. Since $x \in 1 \circ x$ and $x \circ y \subseteq 1 \circ (x \circ y)$, then $1 \circ x \cap F \neq \emptyset$ and $1 \circ (x \circ y) \cap F \neq \emptyset$. Again since F is $\circ$-reflexive, we have $1 \circ x \subseteq F$ and $1 \circ (x \circ y) \subseteq F$. Considering that F is a positive implicative weak hyper filter, it follows that $y \in 1 \circ y \subseteq F$ and thus $y \in F$. Therefore F is a weak hyper filter. □

Example 4.5. *Consider the hyper BE-algebra from Example 3.5. One can calculate that $F = \{1, a\}$ is $\circ$-reflexive and furthermore F is both a weak hyper filter and an implicative weak hyper filter.*

Remark 4.6. (1) *The condition of the $\circ$-reflexivity in Proposition 4.4 is not necessary in general. Consider (1) of Example 4.2 one can check that $F = \{1, a\}$ is both a weak hyper filter and an implicative weak hyper filter, but it is not $\circ$-reflexive since $1 \circ a \cap F \neq \emptyset$ while $1 \circ a = \{a, b\} \not\subseteq F$.*

(2) *The converse of Proposition 4.4 may not be true. Consider $H = \{1, a, b, c, d, e\}$ and the operation $\circ$ is given by the table:*

∘	1	a	b	c	d	e
1	$\{1,c\}$	$\{a\}$	$\{b\}$	$\{c\}$	$\{d\}$	$\{e\}$
a	$\{1,c\}$	$\{1,c\}$	$\{a\}$	$\{1,c\}$	$\{c\}$	$\{d\}$
b	$\{1,c\}$	$\{1,c\}$	$\{1,c\}$	$\{1,c\}$	$\{c\}$	$\{c\}$
c	$\{1,c\}$	$\{a\}$	$\{b\}$	$\{1,c\}$	$\{a\}$	$\{b\}$
d	$\{1,c\}$	$\{1,c\}$	$\{a\}$	$\{1,c\}$	$\{1,c\}$	$\{a\}$
e	$\{1,c\}$	$\{1,c\}$	$\{1,c\}$	$\{1,c\}$	$\{1,c\}$	$\{1,c\}$

Then $(H, \circ, 1)$ is a hyper BE-algebra [11] and $F = \{1, c\}$ is a $\circ$-reflexive subset. Routine calculation shows that F is a weak hyper filter, but it is not an implicative weak hyper filter since $d \circ (a \circ b) = d \circ a = \{1, c\} \subseteq F$ and $d \circ a = \{1, c\} \subseteq F$ while $d \circ b = \{a\} \not\subseteq F$.

Proposition 4.7. *Suppose that F is a nonempty subset of a R-hyper BE-algebra H. If F is an implicative weak hyper filter, then F is a weak hyper filter.*

Proof. Let $x, y \in H$ such that $x \in F, x \circ y \subseteq F$. Since H is a R-hyper BE-algebra, then $1 \circ x = \{x\} \subseteq F$ and $1 \circ (x \circ y) = x \circ y \subseteq F$. Considering that F is a positive implicative weak hyper filter, it follows that $\{y\} = 1 \circ y \subseteq F$ and thus $y \in F$. Therefore F is a weak hyper filter. □

Example 4.8. *Consider the R-hyper BE-algebra from Example 3.3, it can be checked that $F = \{1, b\}$ is both a weak hyper filter and an implicative weak hyper filter.*

Note that the condition of the R-hyper BE-algebra in Proposition 4.7 is not necessary in general and one can see Example 4.6. Moreover the converse of Proposition 4.7 may not be true and see the following example.

Example 4.9. *Consider the R-hyper BE-algebra from Example 3.10. One can check that $F = \{1\}$ is a weak hyper filter, but it is not an implicative weak hyper filter since $a \circ (a \circ b) \subseteq F$ and $a \circ a \subseteq F$ while $a \circ b = \{a\} \not\subseteq F$.*

A weak hyper filter of a hyper BE-algebra may not be an implicative weak hyper filter and one can see Remark 4.6 (2) and Example 4.9. In what follows, we present some conditions that a weak hyper filter of a hyper BE-algebra becomes an implicative weak hyper filter. First, we introduce the concept of distributive and left-transitive hyper BE-algebras which will be used.

Definition 4.10. *A hyper BE-algebra H is said to be*
  (1) *distributive if $x \circ (y \circ z) \leq (x \circ y) \circ (x \circ z)$ for all $x, y, z \in H$;*
  (2) *left-transitive if $y \circ z \leq (x \circ y) \circ (x \circ z)$ for all $x, y, z \in H$.*

Example 4.11. (1) Consider $H = \{a, b, 1\}$ in which the operation $\circ$ is given as follows:

$\circ$	1	a	b
1	$\{1\}$	$\{a\}$	$\{b\}$
a	$\{1\}$	$\{1, a, b\}$	$\{b\}$
b	$\{1\}$	$\{a, b\}$	$\{1, b\}$

Then $(H, \circ, 1)$ is a hyper BE-algebra[18] and moreover we can check that H is distributive.

(2) Consider the hyper BE-algebra from Example 3.3 one can calculate that H is left-transitive.

Theorem 4.12. *Suppose that H is a distributive hyper BE-algebra and F is a $\circ$-reflexive nonempty subset of H. If F is a weak hyper filter, then F is an implicative weak hyper filter.*

Proof. Let $x \circ (y \circ z) \subseteq F$ and $x \circ y \subseteq F$. Since H is distributive then $x \circ (y \circ z) \leq (x \circ y) \circ (x \circ z)$. It follows from (3) of Proposition 2.8 that $(x \circ y) \circ (x \circ z) \cap F \neq \emptyset$. Since $x \circ y \subseteq F$ then there are $a \in x \circ y \subseteq F, b \in x \circ z$ such that $a \circ b \cap F \neq \emptyset$. Considering that F is $\circ$-reflexive we have $a \circ b \subseteq F$. Again since F is a weak hyper filter, we get $b \in F$. Thus $x \circ z \cap F \neq \emptyset$ and so $x \circ z \subseteq F$. It shows that F is an implicative weak hyper filter. □

Example 4.13. *Consider the hyper BE-algebra H from Example 3.8. It can be checked that H is distributive and $F = \{1, a\}$ is $\circ$-reflexive. Moreover F is both a weak hyper filter and an implicative weak hyper filter.*

Corollary 4.14. *Suppose that H is a distributive hyper BE-algebra and F is a $\circ$-reflexive nonempty subset of H. Then F is an implicative weak hyper filter if and only if F is a weak hyper filter.*

Proof. By Proposition 4.4 and Theorem 4.12. □

In the following, we deliver the ways to determine implicative weak hyper filters.

Theorem 4.15. *Suppose that H is a left-transitive hyper BE-algebra and F is a $\circ$-reflexive weak hyper filter of H. If $z \circ (y \circ (y \circ x)) \cap F \neq \emptyset$ and $z \in F$ imply $y \circ x \cap F \neq \emptyset$ for all $x, y \in H$, then F is an implicative weak hyper filter of H.*

Proof. Let $x \circ (y \circ z) \subseteq F$ and $x \circ y \subseteq F$. Then $x \circ (y \circ z) \cap F \neq \emptyset$. Since H is left-transitive then $x \circ (y \circ z) = y \circ (x \circ z) \leq (x \circ y) \circ (x \circ (x \circ z))$. By (3) of Proposition 2.8 $(x \circ y) \circ (x \circ (x \circ z)) \cap F \neq \emptyset$. By $x \circ y \subseteq F$ it yields that there is $a \in x \circ y \subseteq F$ such that $a \circ (x \circ (x \circ z)) \cap F \neq \emptyset$. By the assumption, we can obtain that $x \circ z \cap F \neq \emptyset$. As F is $\circ$-reflexive we get $x \circ z \subseteq F$. Therefore F is an implicative weak hyper filter. □

By the proof of Theorem 4.15, we have the following result immediately.

Corollary 4.16. *Suppose that H is a left-transitive hyper BE-algebra and F is a $\circ$-reflexive weak hyper filter of H. If $z \circ (y \circ (y \circ x)) \cap F \neq \emptyset$ and $z \in F$ imply $y \circ x \subseteq F$ for all $x, y \in H$, then F is an implicative weak hyper filter.*

In what follows, we deliver a characterization of implicative weak hyper filters.

Theorem 4.17. *Suppose that F is a subset containing 1 of a C-hyper BE-algebra H. Then the following are equivalent:*
(1) F is an implicative weak hyper filter;
(2) For every $a \in H$, $F_a = \{x \in H : a \circ x \subseteq F\}$ is a weak hyper filter.

Proof. (1) $\Rightarrow$ (2) Firstly, since H is a C-hyper BE-algebra, then $a \circ 1 = \{1\} \subseteq F$ and so $1 \in F_a$. Assume that (1) holds. Let $x \circ y \subseteq F_a$ and $x \in F_a$ for any $x, y \in H$. Then $a \circ (x \circ y) \subseteq F$ and $a \circ x \subseteq F$. Since F is an implicative weak hyper filter, we have $a \circ y \subseteq F$, namely, $y \in F_a$. Therefore F_a is a weak hyper filter.

(2) $\Rightarrow$ (1) By hypothesis, $1 \in F$. Assume that (2) holds. Let $x \circ (y \circ z) \subseteq F$ and $x \circ y \subseteq F$ for any $x, y, z \in H$. Then $y \circ z \subseteq F_x$ and $y \in F_x$. Since F_x is a weak hyper filter, we can obtain $z \in F_x$, namely, $x \circ z \subseteq F$. Therefore F is an implicative weak hyper filter. □

5 Obstinate weak hyper filters

In this section, we introduce obstinate weak hyper filters in hyper BE-algebras, and mainly discuss the relationships among obstinate weak hyper filters, positive implicative weak hyper filters and maximal weak hyper filters in hyper BE-algebras.

Definition 5.1. *A weak hyper filter F is said to be an obstinate weak hyper filter of H if $x, y \notin F$ implies $x \circ y \subseteq F$ and $y \circ x \subseteq F$ for all $x, y \in H$.*

Example 5.2. *Consider Remark 3.9 (2) one can check that $F = \{1, a\}$ is an obstinate weak hyper filter and $G = \{1\}$ is not an obstinate weak hyper filter since $a, b \notin F$ but $a \circ b = \{1, b\} \not\subseteq F$.*

In general, not every positive implicative weak hyper filter of a hyper BE-algebra is an obstinate weak hyper filter and vice versa.

Example 5.3. (1) *Consider Remark 3.6 (3) $F = \{1, a\}$ is both a weak hyper filter and a positive implicative weak hyper filter of H, but it is not an obstinate weak hyper filter of H since $b \circ c = \{c\} \nsubseteq F$ whence $b, c \notin F$.*

(2) *Consider $H = \{a, b, 1\}$ and the operation $\circ$ is given as follows:*

$\circ$	1	a	b
1	$\{1\}$	$\{a\}$	$\{b\}$
a	$\{1\}$	$\{1, a\}$	$\{1, b\}$
b	$\{1\}$	$\{b\}$	$\{1\}$

Then one can check $(H, \circ, 1)$ is a hyper BE-algebras and $F = \{1, a\}$ is an obstinate weak hyper filter of H, but it is not a positive implicative weak hyper filter of H since $a \in F$ and $a \circ ((b \circ a) \circ b) = \{1\} \subseteq F$ while $b \notin F$.

The following theorem provides a condition that an obstinate weak hyper filter becomes a positive implicative weak hyper filter.

Theorem 5.4. *Suppose that F is a $\circ$-reflexive nonempty subset of H. If F is an obstinate weak hyper filter of H, then F is a positive implicative weak hyper filter.*

Proof. Let $x, y \in H$ such that $(x \circ y) \circ x \subseteq F$. If $x \notin F$, we discuss the following two cases:

Case 1. If $y \in F$, it follows from $y \ll x \circ y$ that $x \circ y \cap F \neq \emptyset$. As F is $\circ$-reflexive, we have $x \circ y \subseteq F$. Combing that $(x \circ y) \circ x \subseteq F$ and F is a weak hyper filter, it yields that $x \in F$, a contradiction.

Case 2. If $y \notin F$, then $x \circ y \subseteq F$ as F is an obstinate weak hyper filter. Similar to the proof of Case 1, we can deduce that $x \in F$, a contradiction.

Therefore $x \in F$ and by Theorem 3.12 F is a positive implicative weak hyper filter. □

Example 5.5. *In Example 3.5 $F = \{1, a, b\}$ is a $\circ$-reflexive positive implicative weak hyper filter of H. One can calculate that F is also an obstinate weak hyper filter of H.*

Remark 5.6. (1) *The condition of the $\circ$-reflexivity from Theorem 5.4 is not necessary in general. Consider Example 3.8, it is not difficult to verify that $F = \{1, b\}$ is both a positive implicative weak hyper filter and an obstinate weak hyper filter, but it is not $\circ$-reflexive since $b \circ a \cap F \neq \emptyset$ while $b \circ a = \{1, a\} \nsubseteq F$.*

(2) *The converse of Theorem 5.4 may not be true. Consider Remark 3.6 (3) it can calculate that $F = \{1, d\}$ is a $\circ$-reflexive subset of H and moreover F is a positive implicative weak hyper filter of H, but it is not an obstinate weak hyper filter since $a, b \notin F$ while $a \circ b = \{b\} \nsubseteq F$.*

In what follows, we introduce the concept of maximal weak hyper filters in hyper BE-algebras in order to explore the relationship between obstinate weak hyper filters and maximal weak hyper filters.

Definition 5.7. *A proper weak hyper filter of H is said to be maximal if it is not a proper subset of any proper weak hyper filter of H.*

Consider Example 3.2 (1) it is easy to verify that $F = \{1, a\}$ is a maximal weak hyper filter of H. In what follows, we deliver a characterization of maximal weak hyper filter of H.

Theorem 5.8. *Suppose that F is a proper weak hyper filter of H. Then the following are equivalent:*
  (1) *F is a maximal weak hyper filter;*
  (2) *$H = (F \cup \{x\}]$, where $x \notin F$ for any $x \in H$.*

Proof. (1) $\Rightarrow$ (2) Assume that F is a maximal weak hyper filter. Then $F \subseteq (F \cup \{x\}]$ and $F \neq (F \cup \{x\}]$. By use of $x \notin F$ and the maximality of F we can get that $H = (F \cup \{x\}]$.

(2) $\Rightarrow$ (1) Assume that G is a proper weak hyper filter of H such that $F \subseteq G$ and $F \neq G$. Then there exists $x \in G$ but $x \notin F$. By (2) we can conclude $H = (F \cup \{x\}]$. Since $(F \cup \{x\}] \subseteq G$ it yields that $G = H$. This shows that F is a maximal weak hyper filter. □

Theorem 5.9. *Suppose that H is a distributive hyper BE-algebra and F is a $\circ$-reflexive weak hyper filter of H. Then for any $a \in H$,*
  (1) *$F_a = \{x \in H : a \circ x \subseteq F\}$ is a weak hyper filter of H;*
  (2) *$F_a = (F \cup \{a\}]$, namely, F_a is a weak hyper filter of H which is generated by F and a.*

Proof. (1) Applying $1 \in a \circ 1$ we have $a \circ 1 \cap F \neq \emptyset$ and hence $a \circ 1 \subseteq F$. It results in $1 \in F_a$. Now let $x \in F_a$ and $x \circ y \subseteq F_a$ for any $x, y \in H$. Then $a \circ (x \circ y) \subseteq F$ and $a \circ x \subseteq F$. Again since H is distributive, we have $a \circ (x \circ y) \leq (a \circ x) \circ (a \circ y)$. Since F is a $\circ$-reflexive weak hyper filter, from (1) and (3) of Proposition 2.8 we can obtain $a \circ y \subseteq F$ and thus $y \in F_a$. It shows that F_a is a weak hyper filter.

(2) Since $1 \in a \circ a$ then $a \circ a \cap F \neq \emptyset$ and so $a \circ a \subseteq F$. Hence $a \in F_a$. Let $x \in F$. Since $x \ll a \circ x$ and F is a $\circ$-reflexive weak hyper filter, from (1) of Proposition 2.8 we have $a \circ x \cap F \neq \emptyset$ and further $a \circ x \subseteq F$. Thus $x \in F_a$ and so $F \subseteq F_a$. Now assume that G is a weak hyper filter of H containing F and a. Let $x \in F_a$. Then $a \circ x \subseteq F$ and hence $a \circ x \subseteq G$. Since $a \in G$ it follows that $x \in G$ and thus $F_a \subseteq G$. Therefore $F_a = (F \cup \{a\}]$. □

Applying Theorem 5.9 we can deliver the following result.

Theorem 5.10. *Suppose that H is a distributive hyper BE-algebra and F is a $\circ$-reflexive nonempty subset of H. Then the following are equivalent:*
   (1) *F is an obstinate weak hyper filter;*
   (2) *F is a maximal weak hyper filter.*

Proof. (1) $\Rightarrow$ (2) Assume that (1) holds and G is a weak hyper filter such that $F \subset G$. Then there are $x \in G, x \notin F$ such that $(F \cup \{x\}] \subseteq G$. Let y be an arbitrary element of H. If $y \in F$ then $y \in (F \cup \{x\}] \subseteq G$. If $y \notin F$ then as $x \notin F$ and F is an obstinate weak hyper filter, we have $x \circ y \subseteq F$. By Theorem 5.9 $y \in (F \cup \{x\}]$ and so $H = (F \cup \{x\}]$. Further $G = H$ and therefore F is a maximal weak hyper filter.

(2) $\Rightarrow$ (1) Assume that F is a maximal weak hyper filter of H. Let $x, y \in H$ such that $x, y \notin F$. Then $H = (F \cup \{x\}]$. According to Theorem 5.9 $H = (F \cup \{x\}] = F_x$ and hence $y \in F_x$. It follows that $x \circ y \subseteq F$. Similarly we can get $y \circ x \subseteq F$. Therefore F is an obstinate weak hyper filter. $\square$

The following results charify the relationship among obstinate weak hyper filters, maximal weak hyper filters, positive weak hyper filters and implicative weak hyper filters in hyper BE-algebras.

Theorem 5.11. *Suppose that H is a distributive hyper BE-algebra and F is a $\circ$-reflexive hyper subalgebra of H. Then the following are equivalent:*
   (1) *F is a maximal and positive implicative weak hyper filter;*
   (2) *F is a maximal and implicative weak hyper filter;*
   (3) *F is an obstinate weak hyper filter.*

Proof. (1) $\Rightarrow$ (2) Assume that F is positive implicative weak hyper filter. Then by Proposition 3.4 F is a weak hyper filter. Let $x \circ (y \circ z) \subseteq F$ and $x \circ y \subseteq F$. Since $x \circ (y \circ z) \leq (x \circ y) \circ (x \circ z)$, it follows from (3) of Proposition 2.8 that $(x \circ y) \circ (x \circ z) \cap F \neq \emptyset$. Combing that $x \circ y \subseteq F$, we have $x \circ z \cap F \neq \emptyset$ and hence $x \circ z \subseteq F$. It implies that F is an implicative weak hyper filter.

(2) $\Rightarrow$ (3) It can be seen by Theorem 5.10.

(3) $\Rightarrow$ (1) Assume that F is an obstinate hyper filter of H. Then by Theorem 5.10 we know that F is a maximal weak hyper filter of H. Again using Theorem 5.4 we can get that F is a positive implicative weak hyper filter. Therefore (1) holds. $\square$

Assume that F is a positive implicative weak hyper filter of H. It follows from Proposition 3.7 that F is a weak hyper filter. Based on this fact, similar to the proof of Theorem 5.11 the following corollary can be acquired immediately.

Corollary 5.12. *Suppose that H is a distributive D-hyper (C-hyper) BE-algebra and F is a ∘-reflexive nonempty subset of H. Then the following are equivalent:*
  (1) *F is a maximal and positive implicative weak hyper filter;*
  (2) *F is a maximal and implicative weak hyper filter;*
  (3) *F is an obstinate weak hyper filter.*

Under the conditions of Theorem 5.11 or Corollary 5.12, we give a diagram to summary the relationships among different types of weak hyper filters in hyper BE-algebras. For simplicity, we'll abbreviate maximal weak hyper filters, obstinate weak hyper filters, positive implicative weak hyper filters and implicative weak hyper filters as MF, OF, PIF and IF, respectively.

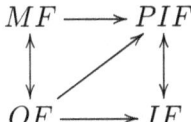

6 Conclusions

Hyper BE-algebras are a generalization of dual hyper BCK-algebras and dual hyper K-algebras. Correspondingly, as a generalization of dual weak hyper ideals in dual hyper BCK-algebras and dual hyper K-algebras, in this paper, we introduce and investigate several types of weak hyper filters in hyper BE-algebras and giving some equivalent characterizations of them especially. Meanwhile, we obtain the relevant theorems that (positive) implicative weak hyper filters become weak hyper filters and find the appropriate conditions of these theorems. Furthermore, we get some important conclusions that state the relationships between maximal weak hyper filters, positive implicative weak hyper filters and obstinate weak hyper filters. On the basis of this study, we can consider fuzzy weak hyper filters and the relations among them, and furthermore investigate the falling shadow theory in hyper BE-algebras, for example, the relationship between falling fuzzy weak hyper filters and falling fuzzy implicative weak hyper filters.

References

[1] R. A. Borzooei, M. Bakhshi and O. Zahiri. Filter theory on hyper residuated lattices. *Quasigroups and Related Systems*, 22(1): 33–50, 2014.

[2] R. A. Borzooei and M. Bakhshi. On positive implicative hyper BCK-ideals. *Scientiae Mathematicae Japonicae Online*, 9: 303–314, 2003.

[3] R. A. Borzooei, M. A. Kologani and M. A. Hashemi. Filter theory on hyper equality algebras. *Soft Computing*, 25(11): 7257–7269, 2021.

[4] R. A. Borzooei, B. G. Saffar and R. Ameri. On hyper EQ-algebras. *Italian Journal of Pure and Applied Mathematics*, 31: 77–96, 2013.

[5] X. Y. Cheng and X. L. Xin. Filter theory on hyper BE-algebras. *Italian Journal of Pure and Applied Mathematics*, 35: 509–526, 2015.

[6] X. Y. Cheng, X. L. Xin and Y. B. Jun. Hyper equality Algebras. *Quantitative Logic and Soft Computing 2016*, Springer, Cham, 415–428, 2017,

[7] X. Y. Cheng, X. L. Xin and Y. W. Yang. Deductive systems in hyper EQ-algebras. *Journal of Mathematical Research with Applications*, 37(2): 183–193, 2017.

[8] P. Corsini and V. Leoreanu. *Applications of Hyperstructure Theory*. Springer Science & Business Media, 2013.

[9] B. Davvaz, A. Dehghan Nezhad and A. Benvidi. Chemical hyperalgebra: Dismutation reactions. *Match-Communications in Mathematical and Computer Chemistry*, 67(1): 55–63, 2012.

[10] B. Davvaz, A. Dehghan Nezhad and M. M. Heidari. Inheritance examples of algebraic hyperstructures. *Information Sciences*, 224: 180–187, 2013.

[11] M. Hamidi, A. Rezaei and A. Borumand Saeid. δ-relation on dual hyper K-algebras. *Journal of Intelligent & Fuzzy Systems*, 29: 1889–1900, 2015.

[12] Y. B. Jun, M. M. Zahedi, X. L. Xin, et al. On hyper BCK-algebras. *Italian Journal of Pure and Applied Mathematics*, 8: 127–136, 2000.

[13] Y. B. Jun and X. L. Xin. Fuzzy hyper BCK-ideals of hyper BCK-algebras. *Scientiae Mathematicae Japonicae*, 53: 353–360, 2001.

[14] H. S. Kim and Y. H. Kim. On BE-algebras. *Scientiae Mathematicae Japonicae*, 66(1): 113–116, 2007.

[15] F. Marty. Sur une generalization de la notion de group. *The 8th Congress Math, Scandinaves, Stockholm*, 45–49, 1934.

[16] G. Muhiuddin, N. Abughazalah, A. Mahboob et al. Hyperstructure Theory Applied to BF-Algebras. *Symmetry*, 15(5): 1106, 2023.

[17] G. Muhiuddin, H. Harizavi and Y. B. Jun. Bipolar-valued fuzzy soft hyper-BCK ideals in hyper-BCK algebras. *Discret. Math. Algorithms Appl.*, 12: 2050018, 2020.

[18] A. Radfar, A. Rezaei and A. Borumand Saeid. Hyper BE-algebras. *Novi Sad J. Math*, 44(2): 137–147, 2014.

[19] M. S. Rao. Positive implicative filters of BE-algebras. *Annals of Fuzzy Mathematics and Informatics*, 7(2): 263–273, 2014.

[20] A. Borumand Saeid, R. A. Borzooei and M. M. Zahedi. (Weak) implicative hyper K-ideals, *Bull. Korean Math. Soc.*, 40(1): 123–137, 2003.

[21] A. Borumand Saeid and M. M. Zahedi. Maximal and Obstinate Hyper K-ideals. *Scientiae Mathematicae Japonicae*, 60(3): 421–428, 2004.

[22] Y. W. Yang, K. Y. Zhu and X. Y. Cheng. Falling fuzzy hyper deductive systems of hyper

residuated lattices. *Italian Journal of Pure and Applied Mathematics*, 44: 965–985, 2020.

[23] Y. W. Yang, K. Y. Zhu and X. L. Xin. Fuzzy weak hyper deductive systems of hyper equality algebras and their measures. *Journal of Intelligent & Fuzzy Systems*, 38(4): 4415–4429, 2020.

[24] M. M. Zahedi. A review on hyper K-algebras. *Journal of Mathematical Sciences and Informatics*, 1(1): 55–112, 2006.

[25] M. M. Zahedi, A. Borumand Saeid and R. A. Borzooei. Implicative hyper K-algebras. *Czechoslovak Mathematical Journal*, 55: 439–453, 2005.

[26] O. Zahiri, R. A. Borzooei and M. Bakhshi. Quotient hyper residuated lattices. *Quasigroups and Related Systems*, 20(1): 125–138, 2012.

www.ingramcontent.com/pod-product-compliance
Lightning Source LLC
Chambersburg PA
CBHW080453170426
43196CB00016B/2789